Underwater Communication: A Guide for Scuba and Commercial Divers

By
Jess Freeman King

Butte Publications, Inc.
Hillsboro, Oregon, U.S.A.

Design and Illustration: Anita Jones
Cover Design and Illustration: Don Lietzau

Butte Publications, Inc.
Hillsboro, Oregon
U.S.A.

ISBN 1-884362-00-1
Printed in U.S.A.

TABLE OF CONTENTS

Preface

This book is designed as a comprehensive text/manual to be used by both sport and commercial divers as a guide to underwater communication.

Recent popularity of and interest in scuba diving and a continued need for a consistent, "universal" language among commercial divers has ushered in the call for a viable form of communication that might be possible in the underwater environment. During the last decade the sport of scuba diving has mushroomed with participation including both men and women of all ages. There has also been a proliferation in the growth of technology and a resultant need for a sophisticated mode of communication for commercial divers to eliminate barriers in a heretofore, for all practical purposes, sterile language environment.

The inadequate and limited gesture system of diving has through the years been rendered obsolete. The gesture system used by divers has typically focused on a limited number of what might be called "survival signs." The possibilities of a total underwater communicative system have not been explored adequately. This has provided the impetus for the present writing.

The purpose of this book is to provide both the scuba diver and the commercial diver with a reference source which will provide a knowledge of and a means for meaningful underwater communication.

— Jess Freeman King

To Benny who started me on the road and to all of the Deaf friends who through the years have taught me their language and afforded me a glance at their culture.

Chapter 1

FINGERSPELLING AND NUMBERS

Fingerspelling

Fingerspelling, or the manual alphabet, involves the use of certain finger configurations and hand positions to visually represent and correspond with the letters of the written alphabet. This form of underwater communication will be of importance to you, the diver, when you do not know or cannot remember a certain sign for the thought you wish to convey.

The beginning underwater communicator should be aware of the necessity of learning to devote practice time to not only expressive communication skills, but receptive skills as well.

There are two areas of specific importance regarding fingerspelling to which you as an underwater communicator should give particular attention.
These are:

1. The position of the hand when spelling and the movement or flow of the fingers themselves;
2. How to develop receptive fingerspelling skills.

1. Hand position and flow of the fingers

The palm of the hand should never face you. Hold the arm in a comfortable, relaxed position.

Do not concern yourself with speed; your emphasis should be placed on clarity of expression. Individual letters should be made distinctly with a slight, barely discernable pause between words. (Initially, you will experience some difficulty in "reading" fingerspelling executed in this manner. However, with practice your skills should improve.) Do not succumb to the temptation to "jerk," "drop," or "push" letters. Each letter should be made clearly.

2. How to develop receptive fingerspelling skills

A major fallacy of most novices attempting to develop receptive fingerspelling skills is the tendency to study the manual alphabet in alphabetical order, letter-by-letter. This should not be done! The manual alphabet is included in this text solely for reference purposes.

Guillory, in her manual, *Expressive and Receptive Fingerspelling for Hearing Adults*, suggests the use of phonetic combinations. Fingerspelling should be read in short units, or groups of letters, rather than letter-by-letter. Initially, these units might be two or three letter combinations which eventually are expanded into words.

Do not become frustrated! The more you use fingerspelling, the more proficient you will become at "reading" as well as "sending." And, remember, the manual alphabet is used only as a supplement to the signs which you will be using in underwater communication.

Manual Alphabet

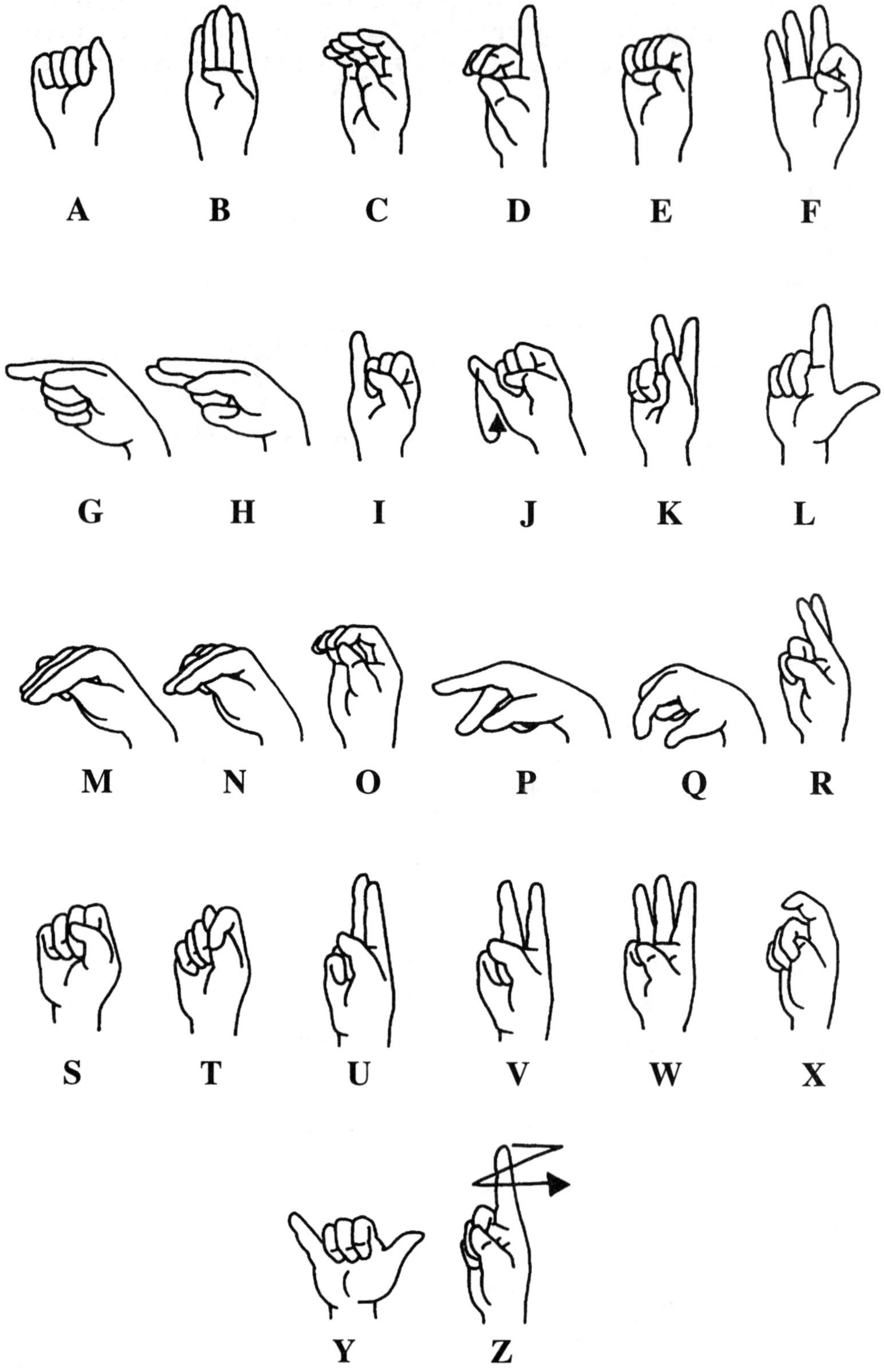

*** The letters are as they would appear to a fellow-diver with whom you are communicating.

Numbers

For the scuba or commercial diver, being able to express depth, pressure, time, or other measures of quantity in a meaningful, explicit manner is of vital importance. Following is a numeral system which you, the underwater communicator, can easily master.

(1)

ONE (1)

The index finger is held straight up (palm facing the diver).

(2)

TWO (2)

The index and middle fingers are held straight up (palm facing the diver).

(3)

THREE (3)

The thumb, index, and middle fingers are held up (palm facing the diver).

(4)

FOUR (4)

The four fingers are held up (palm facing the diver).

(5)

FIVE (5)

All four fingers and the thumb are held up (palm facing the diver).

(6)

SIX (6)

The tip of the thumb and the tip of the little finger touch with the remaining fingers extended (palm facing out).

(7)

SEVEN (7)

The tip of the thumb and the tip of the third finger touch with the remaining fingers extended (palm facing out).

(8)

EIGHT (8)

The tip of the thumb and the tip of the middle finger touch with the remaining fingers extended (palm facing out).

(9)

NINE (9)

The tip of the thumb and the tip of the index finger touch with the remaining fingers extended (palm facing out).

(10)

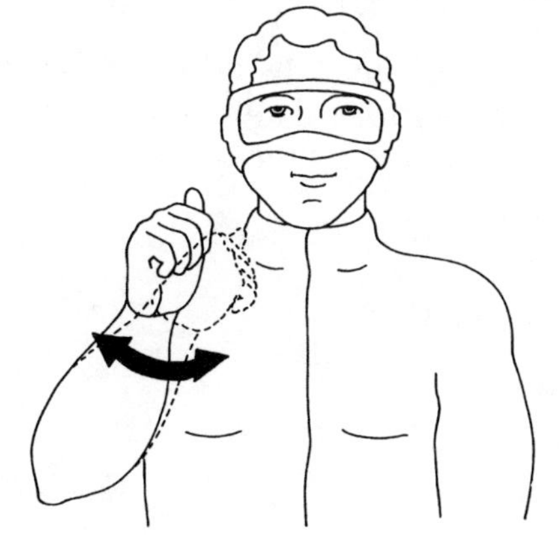

TEN (10)

Make an "A" handshape (palm facing left) with the thumb pointing up and shake it.

(11)

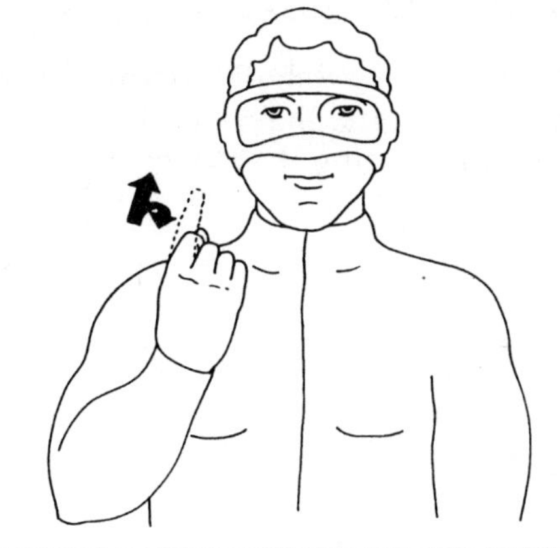

ELEVEN (11)

Make an "A" handshape (palm facing the diver, index and thumbtip touching); flick the index finger up.

(12)

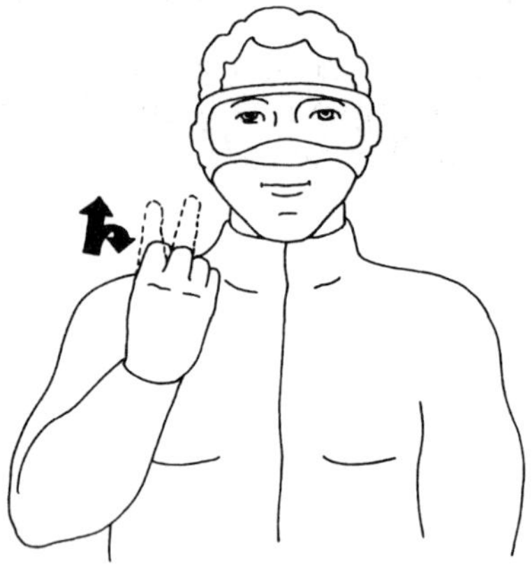

TWELVE (12)

Make an "A" handshape (palm facing the diver, middle and index fingertips touching the thumbtip); simultaneously flick the index and middle fingers up.

(13)

THIRTEEN (13)

Make an "A" handshape (palm facing the diver); immediately make a "3" handshape (palm facing the diver).

(14)

FOURTEEN (14)

Make an "A" handshape (palm facing the diver); now immediately make a "4" handshape (palm facing the diver).

(15)

FIFTEEN (15)

Make an "A" handshape (palm facing the diver); now immediately make a "5" handshape (palm facing the diver).

SIXTEEN (16)

Make an "A" handshape (palm facing the diver); now immediately make a "6" handshape (palm facing out).

SEVENTEEN (17)

Make an "A" handshape (palm facing the diver); now immediately make a "7" handshape (palm facing out).

EIGHTEEN (18)

Make an "A" handshape (palm facing the diver); now immediately make an "8" handshape (palm facing out).

NINETEEN (19)

Make an "A" handshape (palm facing the diver); now immediately make a "9" handshape (palm facing out).

(20)

TWENTY(20)

With the other fingers remaining closed, tap the index finger twice on the thumbtip.

(21)

TWENTY-ONE (21)

Make an "L" handshape (palm facing left) with the index finger pointing forward; wiggle the thumb back and forth several times.

(22)

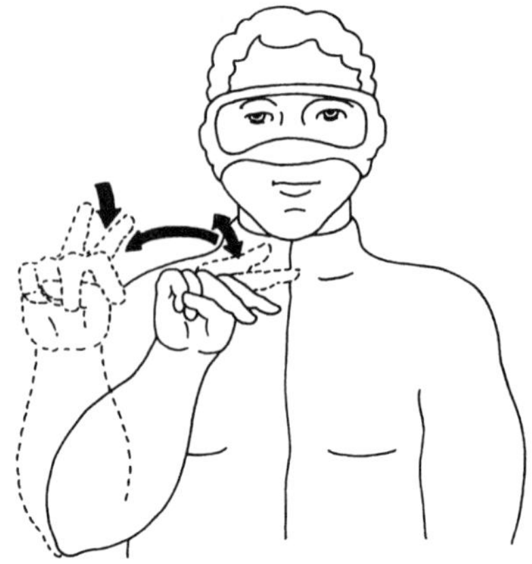

TWENTY-TWO (22)

Make a "2" handshape (palm down); now move the "2" hand in an arc to the right (palm remains down).

(23)

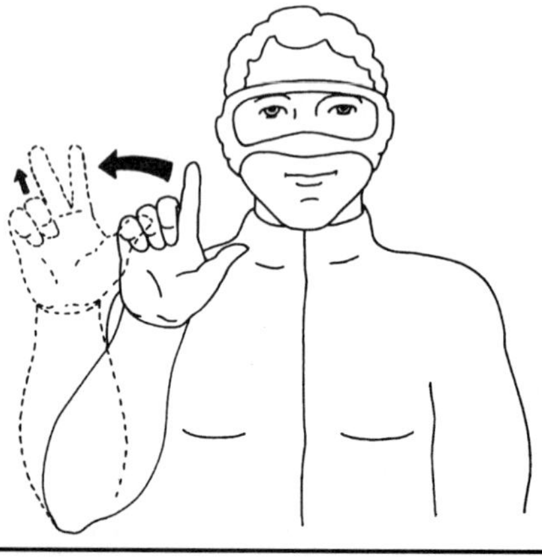

TWENTY-THREE (23)

Make an "L" handshape (palm out); move it to the right forming a "3" handshape.

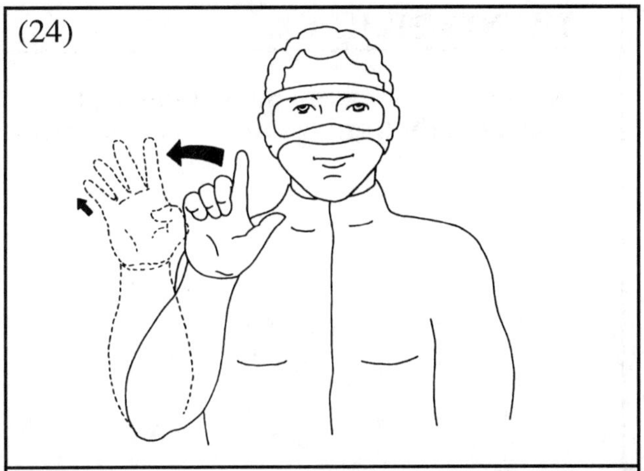

TWENTY-FOUR (24)

Make an "L" handshape (palm out); move it to the right forming a "4" handshape.

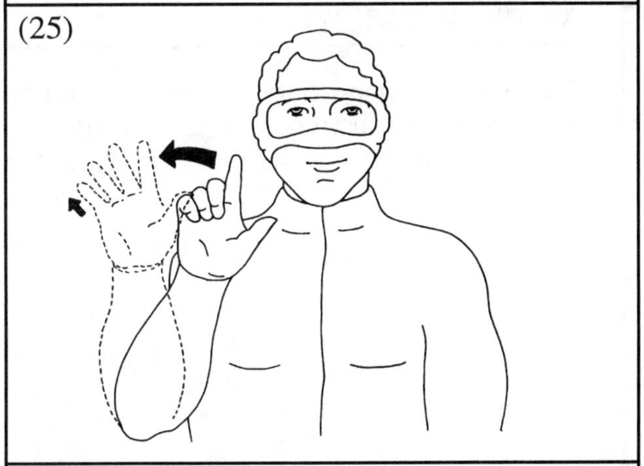

TWENTY-FIVE (25)

Make an "L" handshape (palm out); move it to the right forming a "5" handshape.

TWENTY-SIX (26)

Make an "L" handshape (palm out); move it to the right forming a "6" handshape.

TWENTY-SEVEN (27)

Make an "L" handshape (palm out); move it to the right forming a "7" handshape.

(28)

TWENTY-EIGHT (28)

Make an "L" handshape (palm out); move it to the right forming an "8" handshape.

(29)

TWENTY-NINE (29)

Make an "L" handshape (palm out); move it to the right forming a "9" handshape.

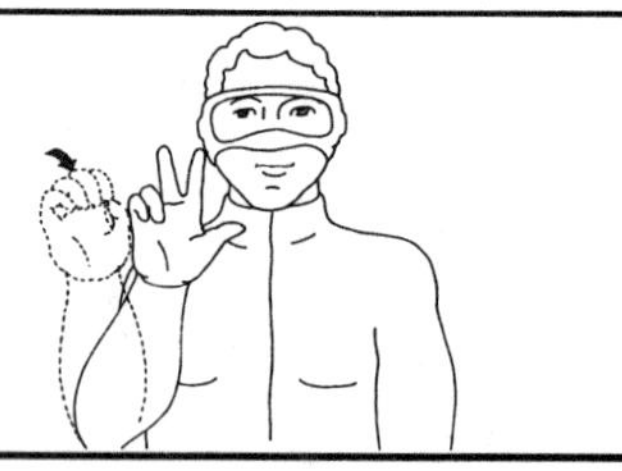

THIRTY

NOTE: 30, 40, 50, 60, 70, 80 and 90 are represented by the base hand positions plus the "0" hand position.

FORTY

SEVENTY

FIFTY

EIGHTY

SIXTY

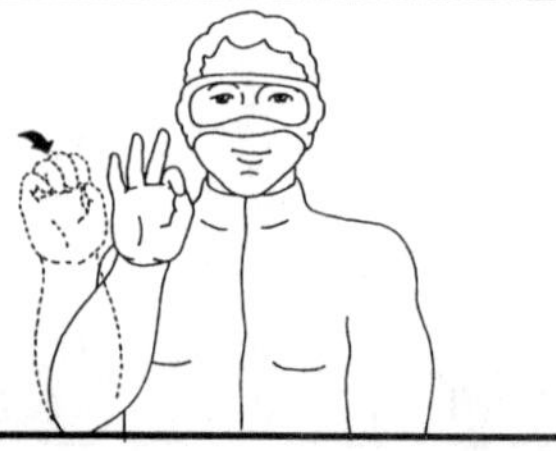

NINETY

(30)

ONE HUNDRED (100)

Make a "1" handshape (palm out); immediately move into a "C" handshape. "C" represents the Roman numeral which stands for 100 units.

(31)

ONE THOUSAND (1,000)

Make a "1" handshape (palm out); immediately make an "M" handshape and touch the "M" fingertips on the left palm. "M" symbolizes the Roman numeral which represents 1,000 units.

NOTE: Fractions are made by signing the upper number (numerator), then slightly lowering the hand and signing the lower number (denominator).

Fingerspelling and Number Drills

The following sentences should be fingerspelled as many times as needed until you feel comfortable with expressive fingerspelling. Remember to practice clarity, not speed!

1. Five fishy fellows found the frog.
2. Divers do not drown in downtown Denver.
3. Sand sharks seldom smell skunks.
4. Most marvelous mermaids made men muddle.
5. The snapper snapped the squid squarely.
6. A blue-bottomed boat boasted of bluefish bisque.
7. Willie Walrus walked wobbily on the wharf.
8. The barracuda blindly bumped the blue bottle.
9. Which way did the ray run with the wrong reel?
10. The skate slid softly through the silky surf.
11. Practice these numbers:

a. 27	f. 16,769
b. 276	g. 55
c. 2,276	h. 530
d. 440	i. 9,914
e. 516	j. 5/8, 4/5, 1/2, 3/4

Prologue to Chapters Two through Seven

A Comparison of Signs and Fingerspelling

You, as an underwater communicator, should be able to differentiate between fingerspelling and signs. As explained in Chapter 1, fingerspelling involves the use of certain finger configurations and hand positions that visually represent the letters of the alphabet. Fingerspelling is English-based.

Signs, on the other hand, are not English-based. Signs employ hand symbols, facial expressions, and "body language" to represent ideas and concepts.

Handshapes and Configurations

In the following chapters you will find such terms as "A" handshape; "5" handshape; the left "open hand;" the "S" handshape, etc. used to explain how the hand is shaped to initiate or to complete a certain sign. For example, when the "S" hand is referred to, this indicates that the hand's configuration corresponds to the "S" position as employed in fingerspelling; the "5" hand, a configuration that corresponds to the "5" position as employed when making the number "5;" and so on.

Also, it should be understood that signs may be executed with either the right or left hand. The diver will note that in the description of many signs the right hand is referred to. However, this is only for description's sake.

Use of Signs by the Diver

Signs offer you, the diver, the opportunity to communicate in the submarine environment in a meaningful, easily understood manner. As you learn signs, the limited, archaic gestures of underwater communication are replaced by a highly sophisticated system of conversation which will provide a means of conveying ideas, thoughts, and vocabulary ranging from the number of scales on a mermaid's tail to the color of the sea urchin to the left of the lobster trap by the sunken ship!

You have launched forth on a course which heretofore has been uncharted. Interact with this "underwater language" - use it - become one of the real pioneers in the venture which can revolutionize both scuba and commercial diving.

Chapter 2

FISHY FACTS

(Underwater critters and related items...)

The intent of this chapter is to provide the underwater-communicator with a lexicon of signs which facilitate the identification and naming of creatures in the submarine environment.

(32)

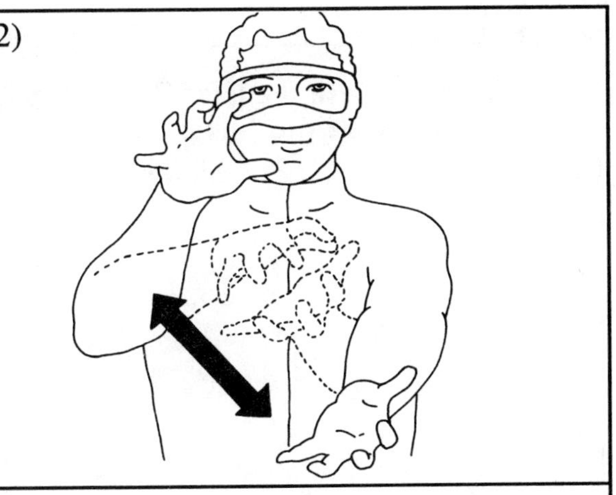

ALLIGATOR

With the heels of the hands touching, the hands are opened and closed several times representing the alligator's mouth opening and closing.

(33)

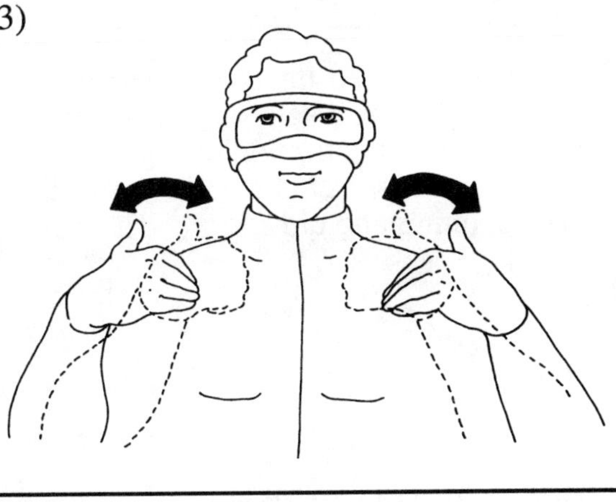

ANIMAL

The fingertips are placed on the chest and the hands are rocked back and forth to indicate the breathing pattern of an animal.

(34)

CLAM

Cup the left hand over the right "C" hand (palm up) and flap the left hand up and down over the "C" hand to simulate the opening and closing of a clam shell.

(35)

CONCH

Both "C" hands are held near the ear as if holding an imaginary conch shell and listening to the "roar of the ocean."

(36)

CORAL

Strike the back of the left "C" hand (palm down) with the little-finger side of the right "C" hand (palm facing left).

(37)

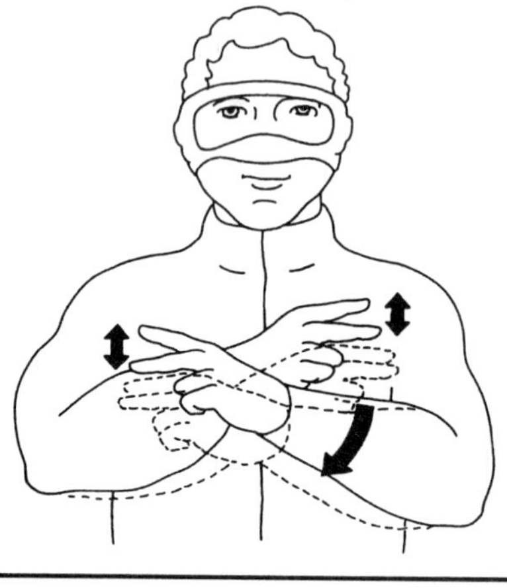

CRAB

Move the crossed-hands (palms down) forward making a scissor-like movement with the index and middle fingers. This movement represents the pincers of the crab opening and closing.

(38)

DOLPHIN / PORPOISE

The "D" hand is moved forward simultaneously dipping up and down to represent the swimming motion of a dolphin.

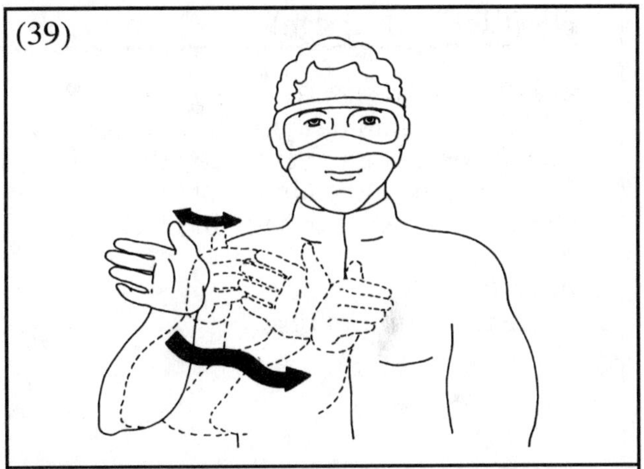

(39)

FISH

The right hand is pointed forward (palm facing left) and then wiggled back and forth as it is moved forward representing the swimming motion of a fish.

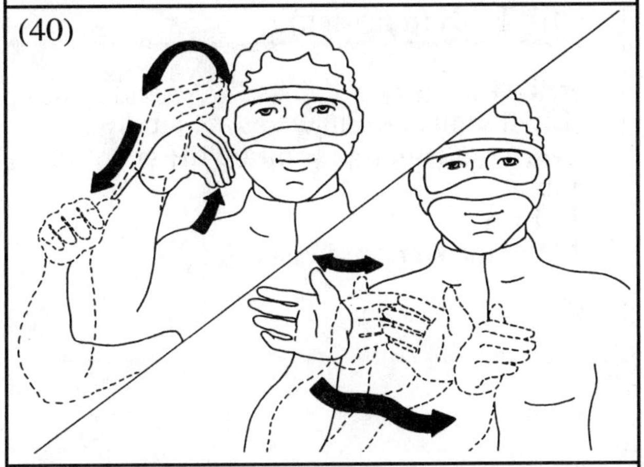

(40)

ANGELFISH (dual sign)

a. Angel
The fingertips of the right "open" hand are touched to the right shoulder and the hand is then arched to the right away from the shoulder indicating the shape of an angel's wing.
b. Fish
The right hand is pointed forward (palm facing left) and then wiggled back and forth as it moves forward representing the swimming motion of a fish.

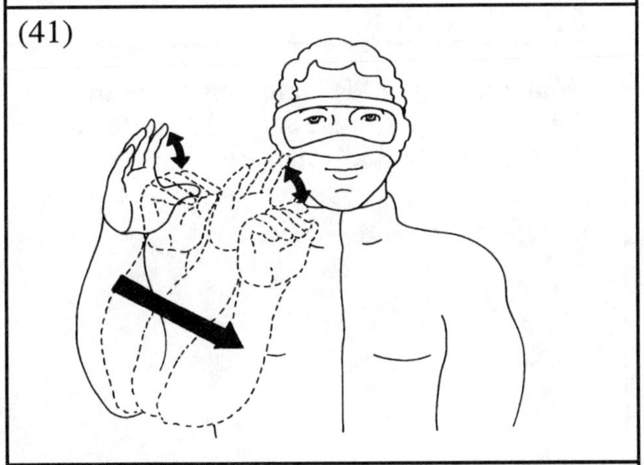

(41)

BARRACUDA

The right open hand (palm down, fingers extended) is moved from right to left in front of the diver as the fingers and thumb snap open and shut several times representing the mouth of a barracuda opening and snapping shut.

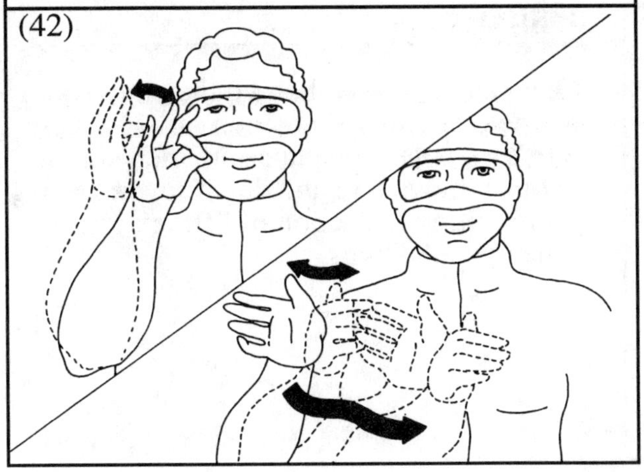

(42)

CATFISH (dual sign)

a. Cat
The "F" hand is touched to the cheek and drawn away several times representing the whiskers on a cat.
b. Fish
Make the sign for "Fish."

(43)

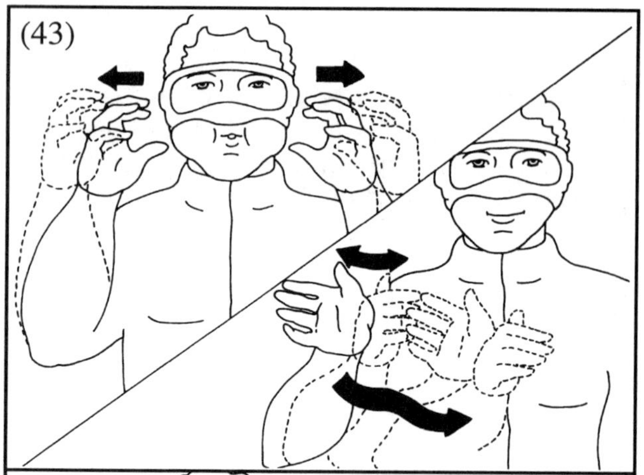

GROUPER (dual sign)

a. Fat
The bent fingers of the "5" hands (palms facing cheeks) are drawn slightly away from the face indicating chubby or "fat" cheeks. A grouper usually has a large, puffy face.
b. Fish
Make the sign for "Fish."

(44)

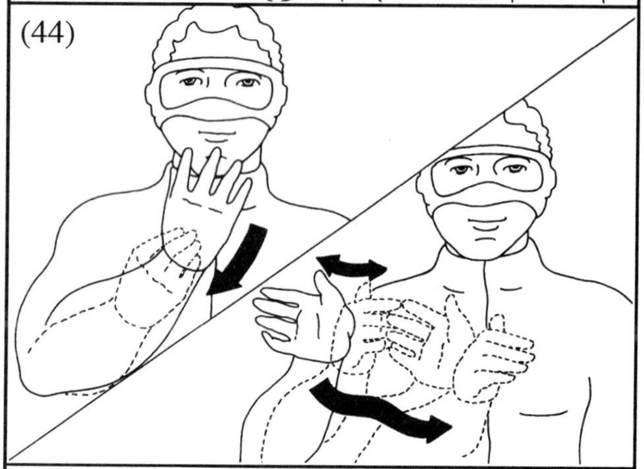

JEWFISH (dual sign)

a. Jew
The thumb and fingertips are stroked down the chin representing the beard of a rabbi.
b. Fish
Make the sign for "Fish."

(45)

RED SNAPPER

Make the letter "R" and then the letter "S." The "R" of course, is the initial for "red" and the "S," snapper.

(46)

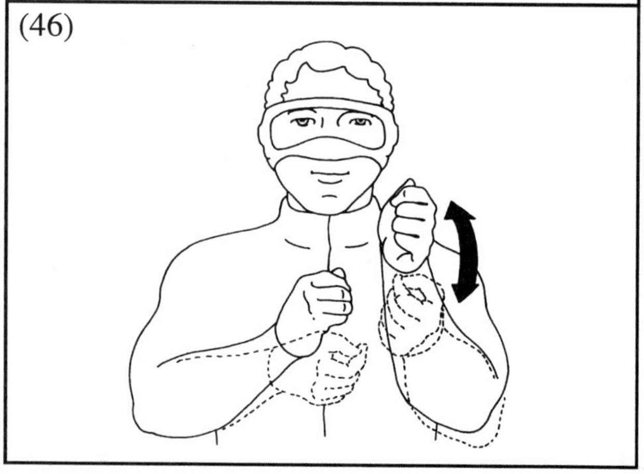

FISHING

One hand is placed behind the other (both are held in the "A" position) as if holding a fishing pole. The hands are then simultaneously moved up and down several times representing the action of lifting a fishing pole up and down.

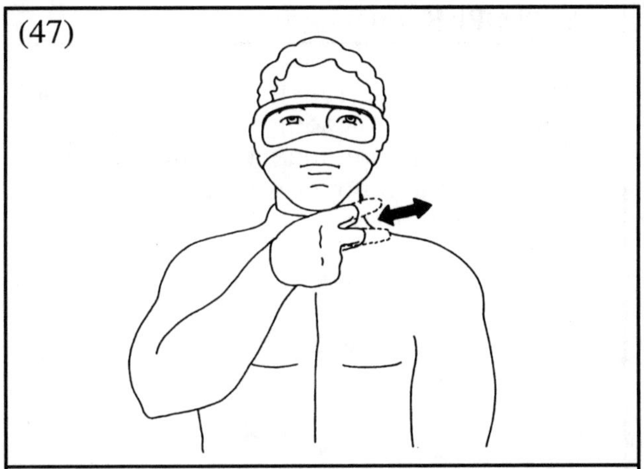

FROG

The “S” hand is placed near the throat; then the index and middle fingers are snapped out several times. This action simulates the croaking action of a frog.

KELP

The hands (palms facing the diver, fingers spread pointing upward) are moved upward in a wavy pattern representing the swaying motion of kelp as it grows toward the surface.

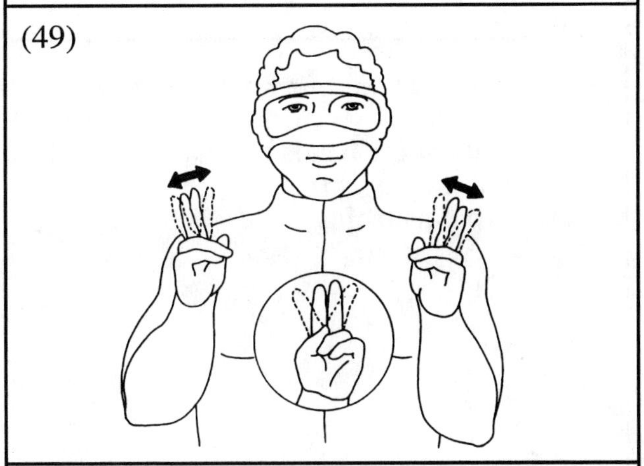

LOBSTER

With the hands held 10"-12" apart (palms facing), make a scissor-like movement with the index and middle fingers of each hand. This movement represents the claws of the lobster opening and closing.

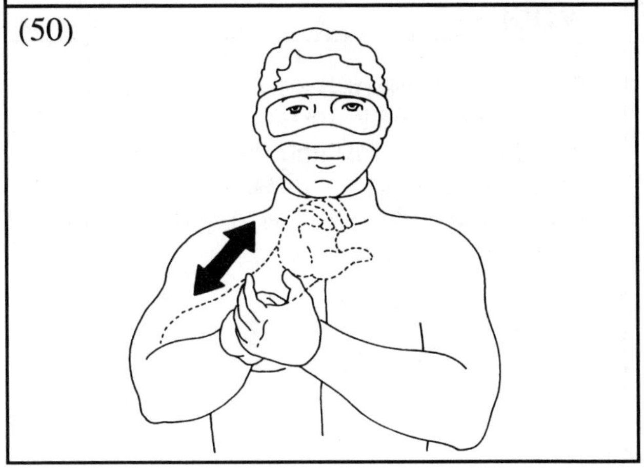

MORAY EEL

With the left “C” hand facing the diver the right “C” hand is thrust upward and forward through the left “C” hand, then abruptly withdrawn. This movement represents the moray eel emerging from a layer of rock or a crevice and then returning to his original position.

(51)

OCTOPUS

The "S" hand is placed on the back of the relaxed, fingers- extended hand. The relaxed, fingers-extended are then wiggled simulating the movement of the tenacles on the octopus.

(52)

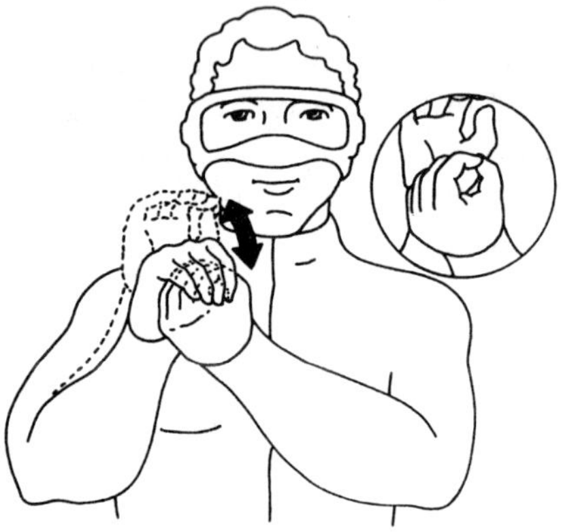

OYSTER

Cup the right hand over the left "O" hand (palm up) and flap the right hand up and down over the "O" hand to simulate the opening and closing of an oyster shell.

(53)

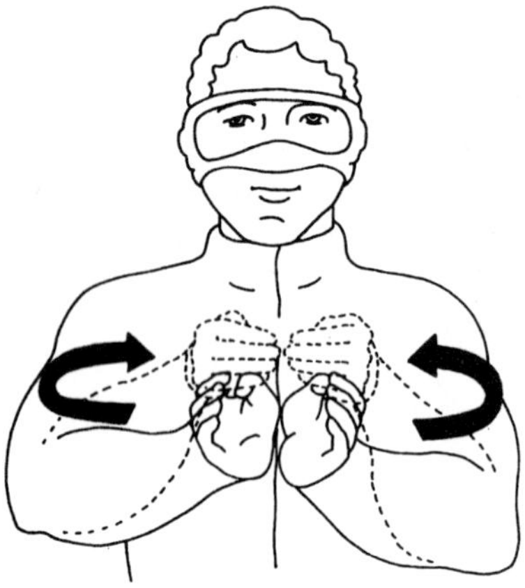

RAY

(can represent any member of the ray family)
The heels of the hands touch and then the shape of the ray is traced with the open hands (palms facing).

(54)

ROCK

Strike the back of the left "S" hand (palm down) with the knuckles of the right "S" hand (palm down) and then indicate the size of the rock.

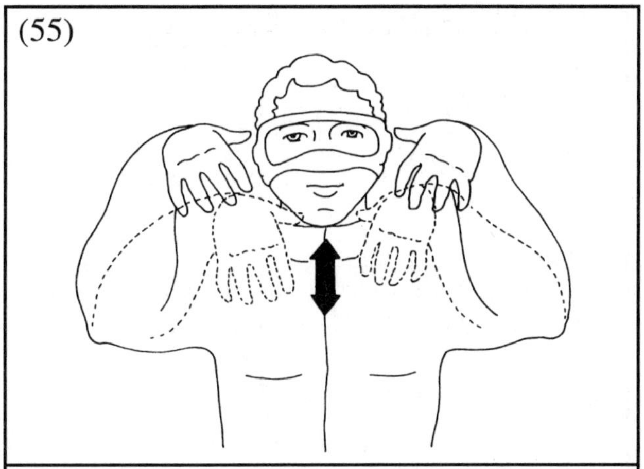

SEAWEED

The hands (palms facing the diver, fingers extended, pointing downward) are moved up and down several times to indicate seaweed bobbing up and down in the water.

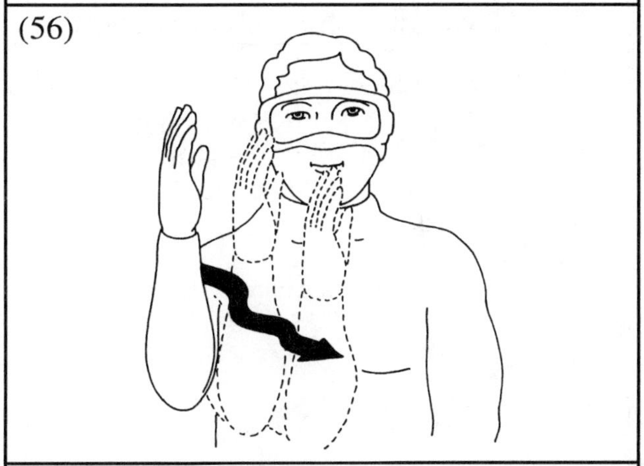

SHARK

The right hand is held (palm facing left) with the fingertips pointing upward. This position represents the shark's dorsal fin. The hand is then moved forward in a zigzag pattern simulating the swimming motion of a shark.

BULL SHARK (dual sign)

a. Bull
The "Y" hand is placed palm out on the forehead. The thumb and little finger represent the bull's horns.
b. Shark
Sign the word for "Shark."

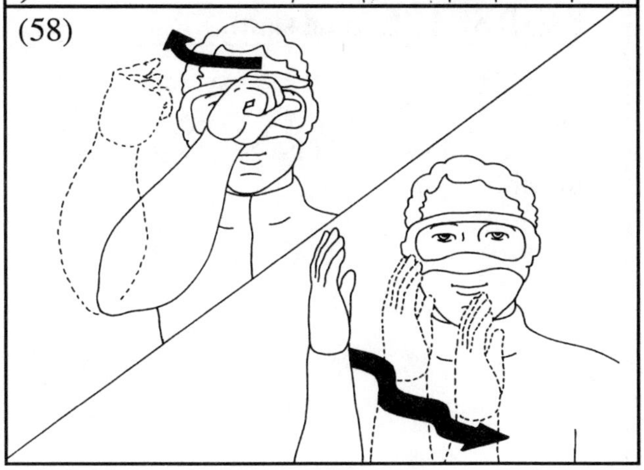

GRAY SHARK (dual sign)

a. Gray
The right "G" hand is moved from left to right across the forehead.
b. Shark
Sign the word "Shark."

(59)

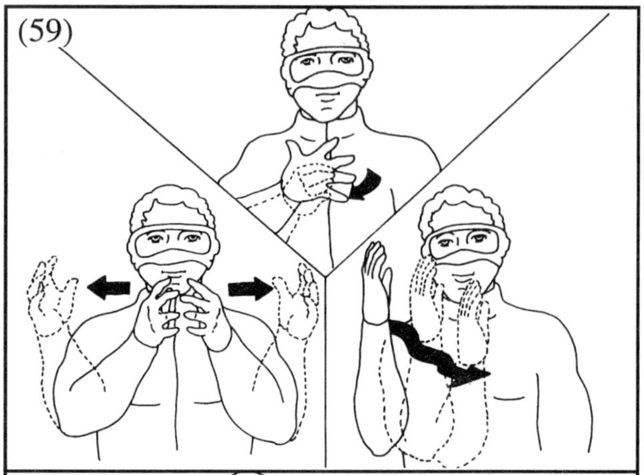

GREAT WHITE SHARK (tri-sign)

a. Great
The "5" hands touch palm-to-palm and then are drawn apart indicating the size of the great white.
b. White
The thumbtip and fingertips touch the chest and are then drawn forward with fingertips and thumbtips touching.
c. Shark
The sign "Shark" is given.

(60)

HAMMERHEAD SHARK (dual sign)

a. Hammerhead
The "S" position hands (palm out) on both sides of the head are moved out away from the head to indicate the shape of the shark's head.
b. Shark
The sign "Shark" is given.

(61)

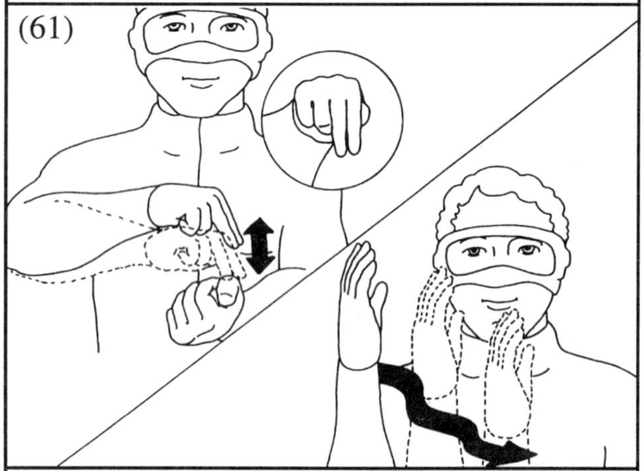

NURSE SHARK (dual sign)

a. Nurse
The "N" fingers of the right hand are tapped twice on the pulse of the left wrist.
b. Shark
Sign the word "Shark."

(62)

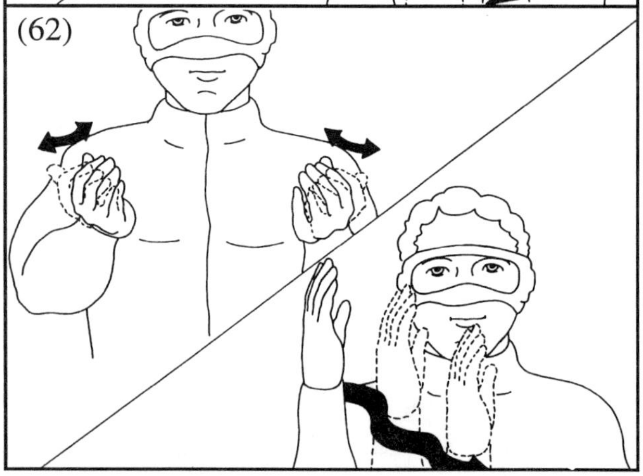

SAND SHARK (dual sign)

a. Sand
With the palms turned upward, the thumbs are rubbed back and forth across the fingertips several times.
b. Shark
Sign "Shark."

(63)

TIGER SHARK (dual sign)

a. Tiger
Both "5" hands (fingers bent, pointing toward the head; palms facing the diver) are drawn outward from either side of the area of the nose indicating the tiger's stripes.
b. Shark
Sign the word "Shark."

(64)

WHITE-TIPPED SHARK (tri-sign)

a. White
Sign "white" as indicated for great white shark.
b. Tipped
Touch middle fingertip of the left "shark" hand with the index finger of the right hand.
c. Shark
Sign the word "Shark."

(65)

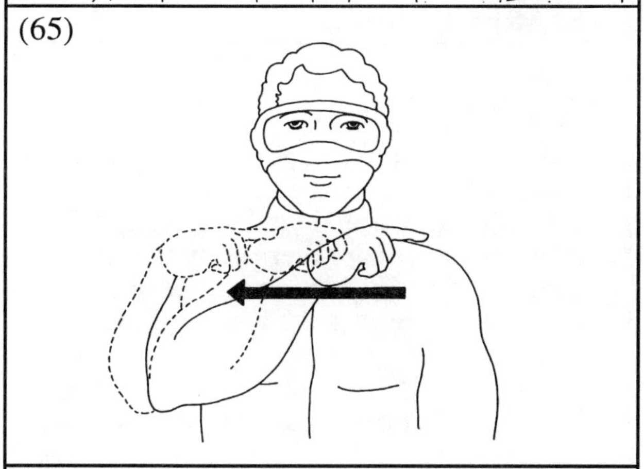

SHRIMP

The right "X" hand (palm down) is moved from left to right in front of the diver alternately extending and contracting the index finger to represent the action of a shrimp swimming through the water.

(66)

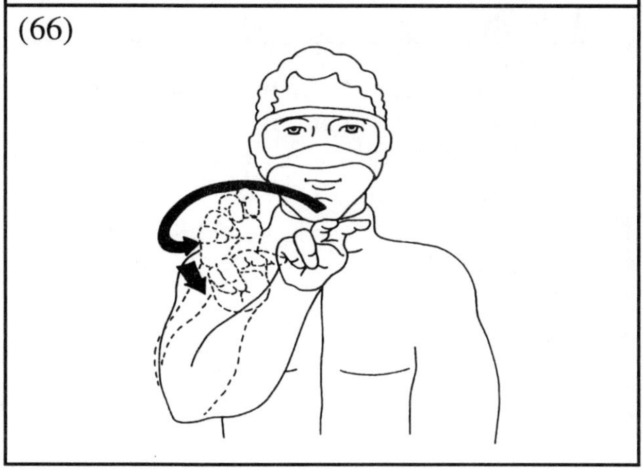

SNAKE

The wrist is rotated in a circular motion as the "V" fingers thrust forward representing the striking motion of a snake.

(67)

SPONGE

The "S" hands (heels of the hands touching) alternately open to a "5" position and then close several times representing the action of a sponge as it takes in water, then squeezes it out.

(68)

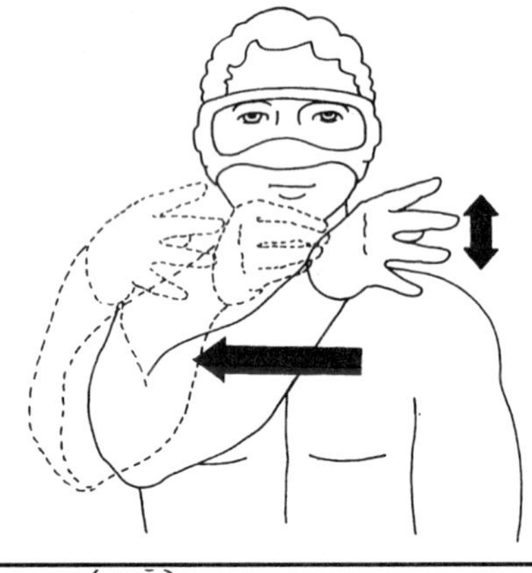

SQUID

The right hand (palm facing left) is drawn from left to right in front of the diver, fingertips and thumbtips alternately touching representing the swimming action of a squid.

(69)

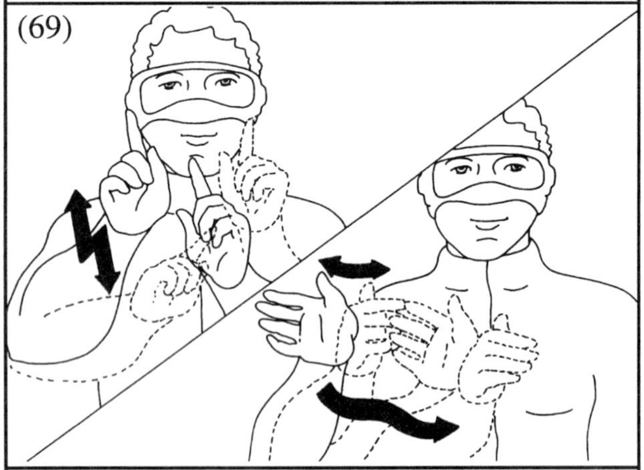

STARFISH (dual sign)

a. Star
With the palms facing forward and both index fingers pointing up, the side of the right index finger strikes the side of the left index finger and arcs to the right. The movement is then reversed by the left index finger. This action is repeated several times and represents the twinkling of a star.
b. Fish
Use the sign previously learned for "Fish."

(70)

URCHIN

The bent-fingered "5" hand (palm up) is moved sharply up and down twice indicating the spines of the sea urchin.

Chapter 3

ON THE BOAT

The signs in this chapter by and large pertain to the items and people who are deckside. However, there will be times when the discussion of these terms will be in the underwater setting. With this thought in mind, the following signs are presented.

(71)

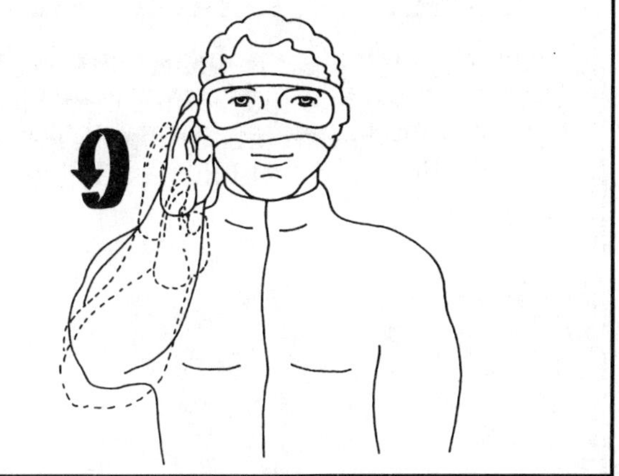

BEER

The right "B" hand is circled on the right cheek indicating the rosy cheeks of a beer drinker.

(72)

BEER (alternative)

The thumb of the right "Y" hand is touched to the lips as if holding a mug of beer and taking a swallow.

(73)

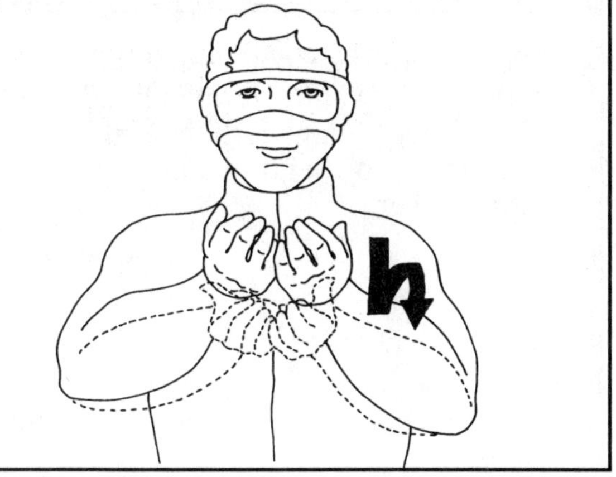

BOAT

The little finger edges of the open hands are touched together to form the shape of a boat. The hands are then moved forward in a wavy, up and down motion representing the action of a boat in the water.

(74)

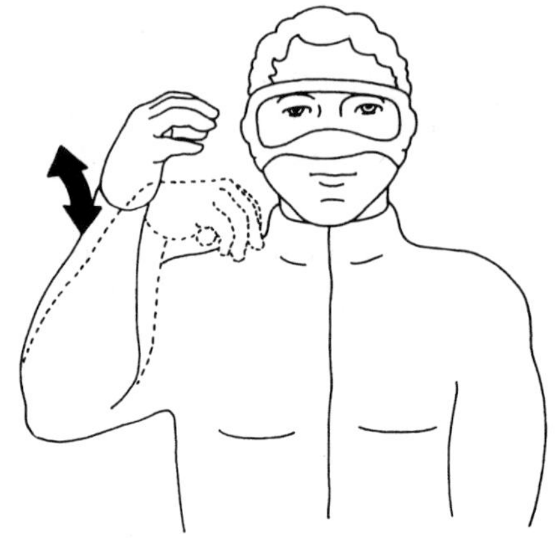

CAPTAIN

The fingertips of the right curved "5" hand are tapped on the right shoulder twice representing the epaulets which were once worn by those in positions of power or authority.

(75)

COFFEE

The right "S" hand is placed on top of the left "S" hand and moved in a circular, grinding manner representing coffee being ground in an old-fashioned coffee grinder.

(76)

COLA

The index finger of the right "L" hand is tapped twice on the left bicep representing a "shot in the arm" as a result of having drunk a cola.

(77)

COOK

The back of the right open hand is placed in the palm of the left open hand and flipped over (palm-to-palm) and back several times representing a pancake being flipped over.

(78)

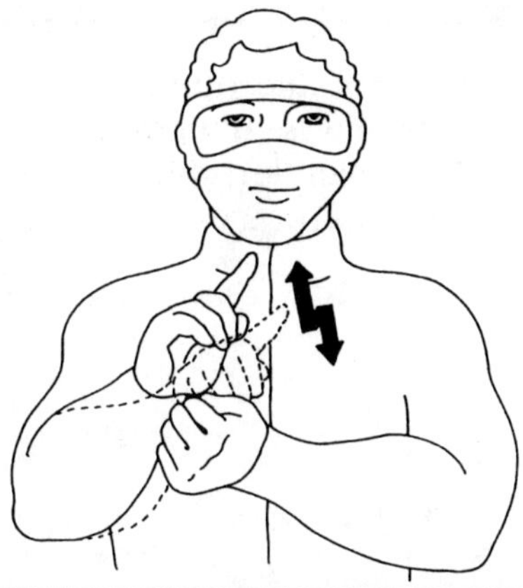

DOCTOR

The thumb of the middle finger of the right "D" hand taps the pulse area of the left wrist representing the taking of the pulse by the doctor.

(79)

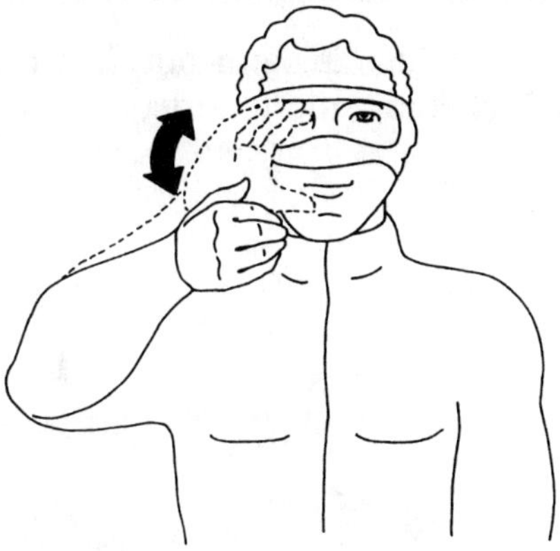

DRINK

The right "C" hand, (palm facing left) is brought to the lips and tilted toward the mouth or regulator. The "C" hand represents a cup of water or other liquid.

(80)

EAT

With the thumbtip touching the fingertips, touch the lips or regulator twice.

(81)

HUNGRY

Draw the "C" hand (palm facing the diver) once down the chest. This action represents the food's passage into the stomach.

(82)

MEDICINE

The tip of the middle finger of the right hand touches and is rubbed in a circular motion in the open left palm. This action represents the medicine being mixed in the palm.

(83)

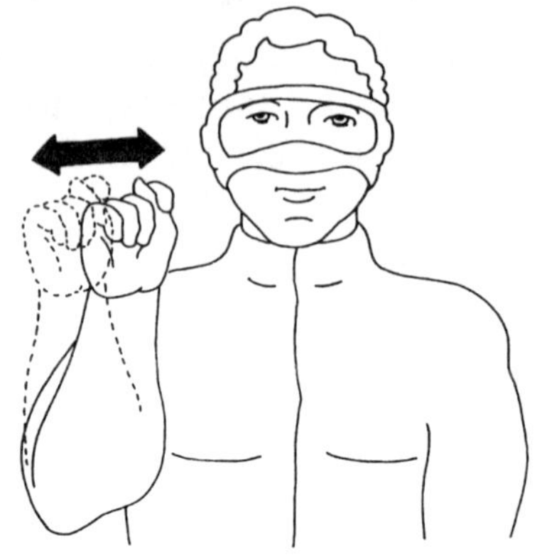

RESTROOM

Hold the "T" hand (palm-out) in front of you and shake it side-to-side.

(84)

SHIP

The right "3" hand is placed in the left hand palm and the hands are moved forward in a wavy up and down motion to simulate the action of a ship in the water. The "3" hand represents the masts on the ship.

(85)

SLEEP

The head is tilted to the side and rested on the open hands held palm to palm. This action represents the head resting on a pillow.

(86)

TEA

The thumb and index finger of the right "F" hand are placed inside of the left "O" hand and stirred around and round. The "F" hand represents a tea bag being stirred in a cup ("O" hand).

(87)

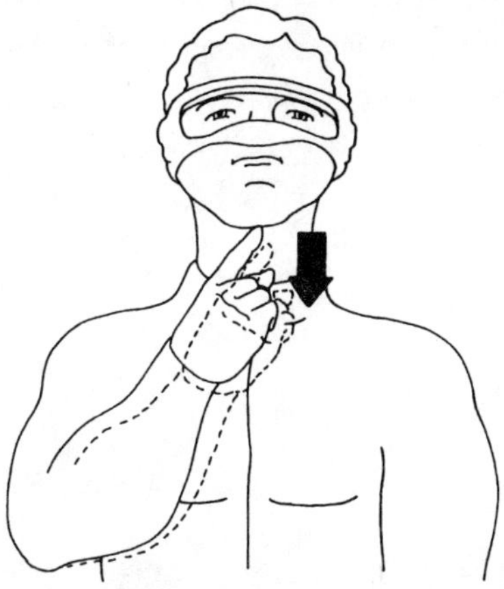

THIRSTY

The index finger (palm facing the diver) is drawn down the throat representing dryness of the throat.

(88)

WINE

The right "W" hand (palm-in) is circled on the right cheek.

Chapter 4

GEAR UP!

The signs in this chapter are of great importance to the underwater communicator. If the diver, for whatever reason, is experiencing a problem equipment-wise while below the surface, it is important that he/she be able to communicate this fact to their fellow divers. It is with this thought in mind that the following signs are included in the text.

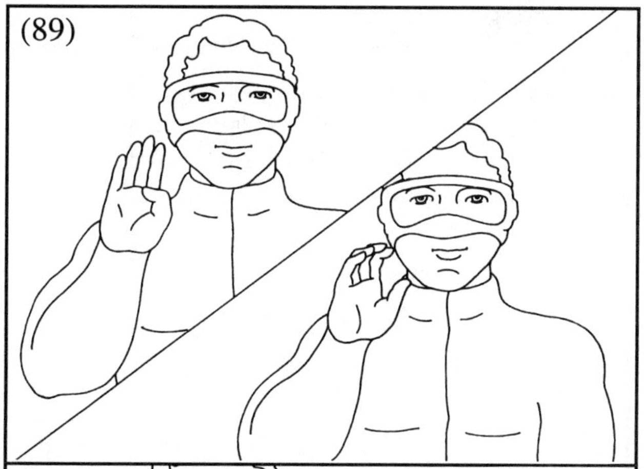

(89)

BUOYANCY COMPENSATOR

The "B" hand position is made, followed by the "C" hand position. (The initials BC, of course, standing for buoyancy compensator.)

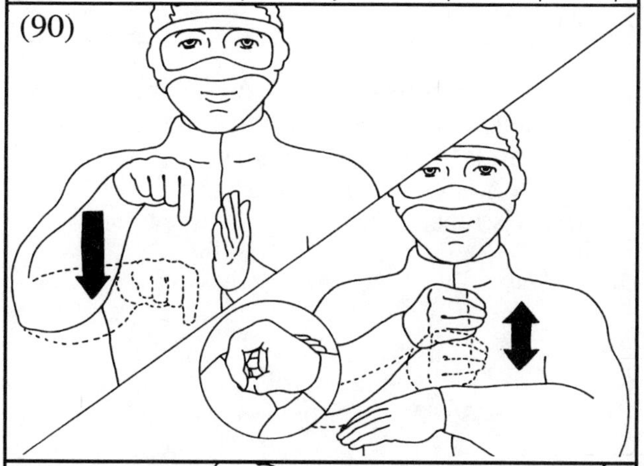

(90)

DEPTH GAUGE (dual sign)

a. Depth
The right index finger is slowly pushed straight down beside the left palm which is facing right. The index finger moving straight down represents depth.
b. Gauge
The right "O" hand is placed, little finger touching, on top of the left wrist indicating the shape of a gauge.

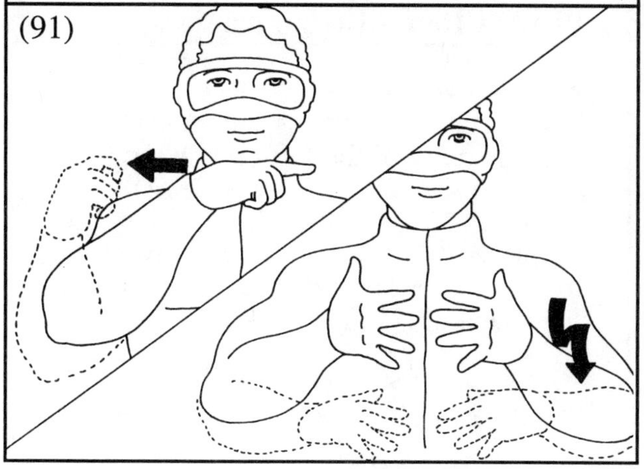

(91)

DRY SUIT (dual sign)

a. Dry
The right "X" hand (palm down) is drawn from left to right along the lips to represent dryness of the lips.
b. Suit
The "5" hands (palms facing the diver) are drawn down the chest two times allowing the thumbs to touch the chest.

(92)

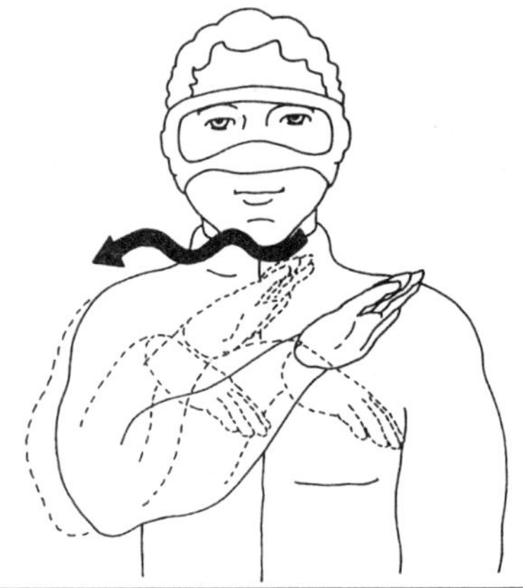

FINS

The "B" hand (palm down) is flipped up and down from the wrist several times to simulate the action of the diver's fin as it propels the diver through the water.

(93)

KNIFE

The index finger of the right hand (palm down) is flicked off the upper part of the index finger of the left hand (palm facing right) to indicate a knife being sharpened.

(94)

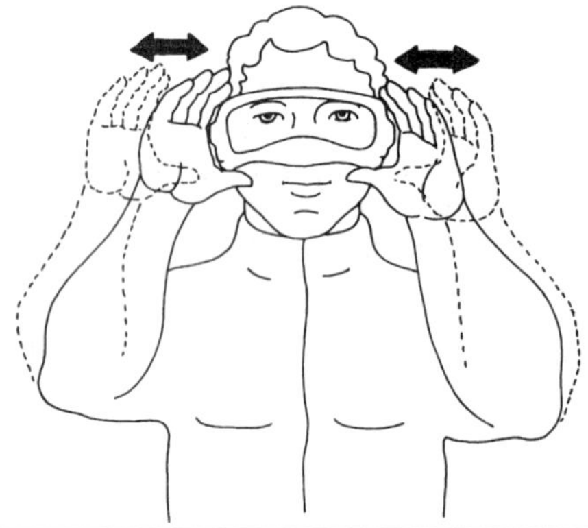

MASK

The "C" hands (palms facing) are placed at the sides of the head with the thumbs touching the cheekbones. This position of the hands represents the shape of the diver's mask.

(95)

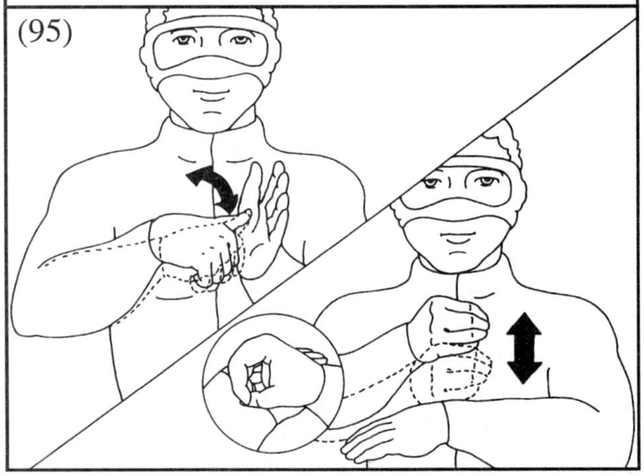

PRESSURE GAUGE (dual sign)

a. Pressure
The thumbtip of the "A" hand is pressed into the palm of the left open hand and twisted twice in a clockwise direction.
b. Gauge
The right "O" hand is placed, little finger touching, on top of the left wrist indicating the shape of a gauge.

(96)

REGULATOR

The "5" hand, (palm facing the diver and fingers bent) is touched twice to the mouth indicating the position of the regulator in the diver's mouth.

(97)

SNORKEL

With the thumb and index finger in a modified "C" position (other fingers closed), trace the shape of a snorkel beginning near the mouth and ending above the head.

(98)

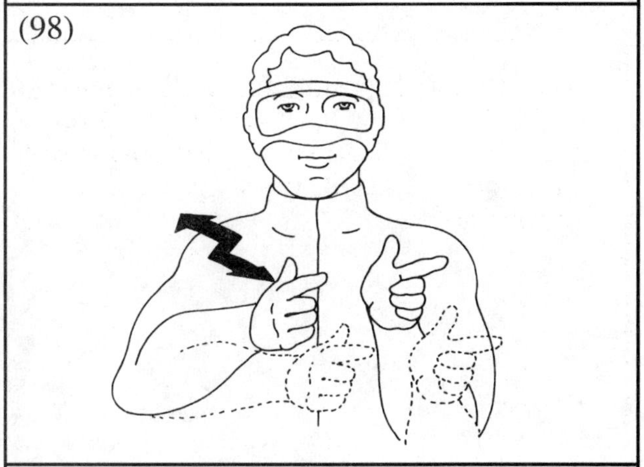

SPEARGUN

The right "L" hand (palm left and index finger pointing forward) is held behind the left "L" hand (palm right and index finger pointing forward); the hands are then moved forward-back-forward to represent the shooting action of the gun and releasing the spear.

(99)

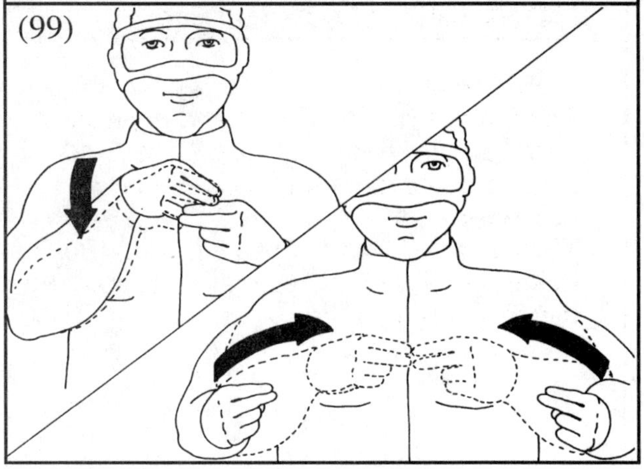

WEIGHT BELT (dual sign)

a. Weight
The middle finger of the right "H" hand (palm facing left shoulder) is placed across the index finger of the left "H" hand (palm facing right shoulder) and rocked back and forth representing a scale being balanced.
b. Belt
The "H" hand fingertips (palms facing in) are overlapped in front of the diver's waist indicating the buckling of a belt.

(100)

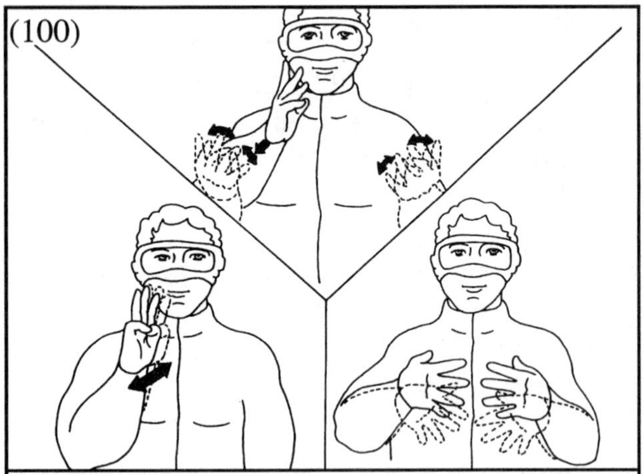

(101)

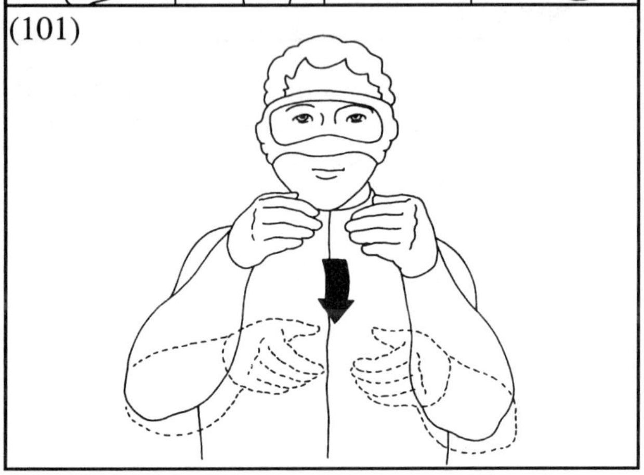

WET SUIT (tri-sign)

a. Water
The index finger of the right "W" hand (palm facing left) is touched to the side of the lips.
b. Wet
Then, the "5" hands are placed palm-up in front of the diver, and the fingertips and thumbtip of each hand touch several times.
c. Suit
The "5" hands (palms facing the diver) are drawn down the chest two times allowing the thumbs to touch the chest.

TANK

The "C" hands (palms facing) are drawn downward in front of the body to represent the cylindrical shape of the tank.

Chapter 5

YOU BET YOUR LIFE!

(Survival Signs)

How many divers have been injured or lost their lives simply because they were not able to adequately and with meaning communicate with their diving partner? With a knowledge of survival signs and proper training this should not happen.

For the diver who has aspirations of becoming an underwater communicator, this is perhaps the one chapter which should take precedence above all the others.

(102)

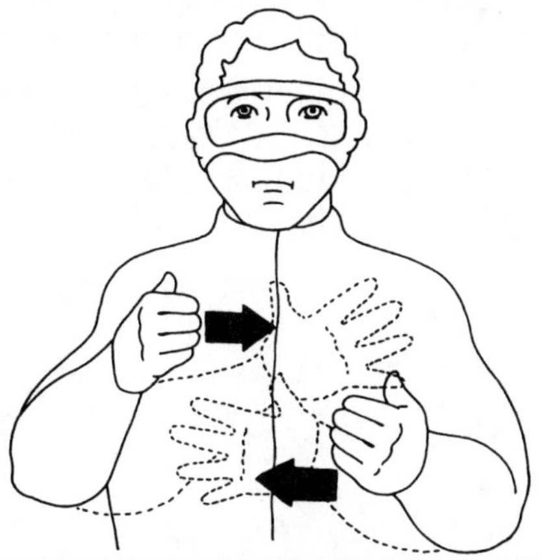

AFRAID

The "A" hands (palms facing the chest) are opened and simultaneously the right hand moves to the diver's left and the left hand moves to the right. This action represents the diver trying to protect himself from what he is "afraid" of.

(103)

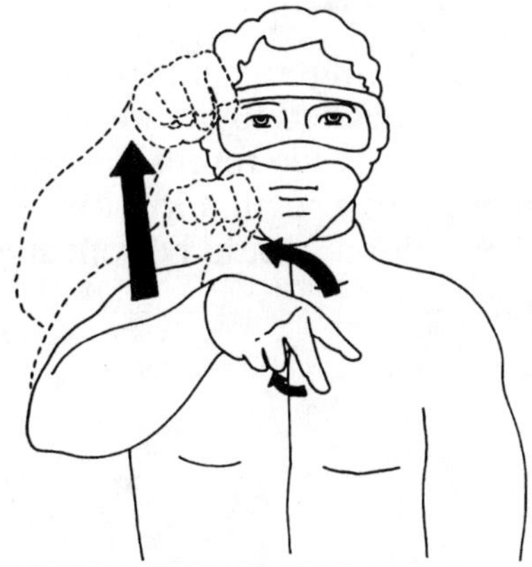

BENDS

The index and middle fingers of the right "V" hand (palm facing the diver) are bent as the hand is drawn upwards. This movement represents the "bending" of the body during ascent.

(104)

BETTER

The fingertips of the right open hand (palm facing the diver) are touched to the regulator and then closed into an "A" position as the hand is drawn up and to the side of the head.

(105)

BLOOD

The right index finger is drawn down the front of the regulator or lips (sign for red); then the right fingertips are wiggled down the back of the left hand to indicate the blood flowing or dripping down.

(106)

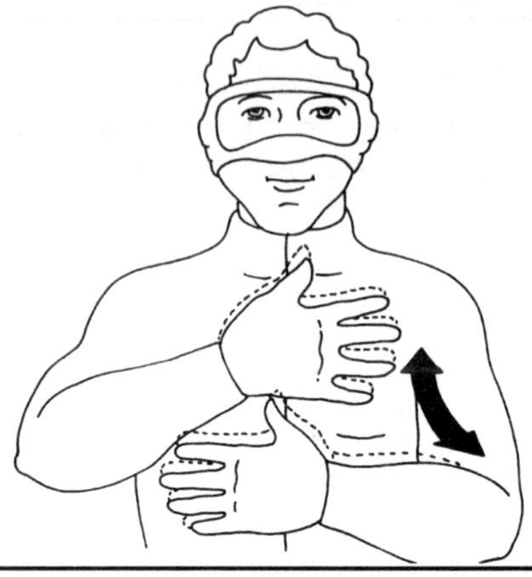

BREATHE

The “5” hands (palms facing the diver) are moved simultaneously on and off the chest several times to indicate the movement of the chest in breathing.

(107)

BUDDY BREATHE

The “5” hand, fingers bent (palm facing the diver) is touched to the mouth and then moved to the outside representing the diver sharing this regulator with another diver. This action should be repeated twice.

(108)

CAREFUL / BE CAREFUL

The right “V” hand (palm facing left) is tapped twice on top of the left “V” hand (palm facing right).

(109)

CARELESS

The right "V" hand (palm facing left) is moved back and forth in front of the eyes several times.

(110)

COMFORTABLE

The right open hand (palm down) is stroked across the back of the left open hand (palm down) from wrist to fingertips; then the left open hand (palm down) is stroked across the back of the right open hand (palm down) from wrist to fingertips.

(111)

COUGH

The thumbtip and fingertips of the right "C" hand are touched to the chest, and the hand is rocked back and forth several times to represent the act of coughing.

(112)

CRAMP

The right "5" hand (palm facing left) is placed above the left "5" hand (palm facing right); now both hands are closed into tight fists representing the muscles contracting in a cramp. Then point to the area of the cramp.

(113)

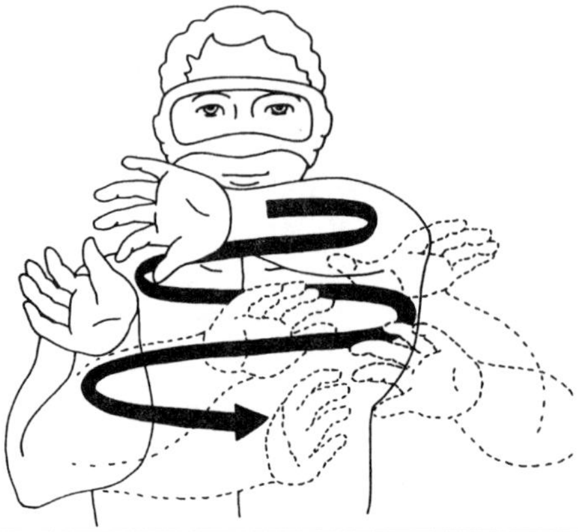

CURRENT

The “5” hands (palms facing left, left hand above right) are swayed from right to left and back several times. This movement represents the flowing of the current.

(114)

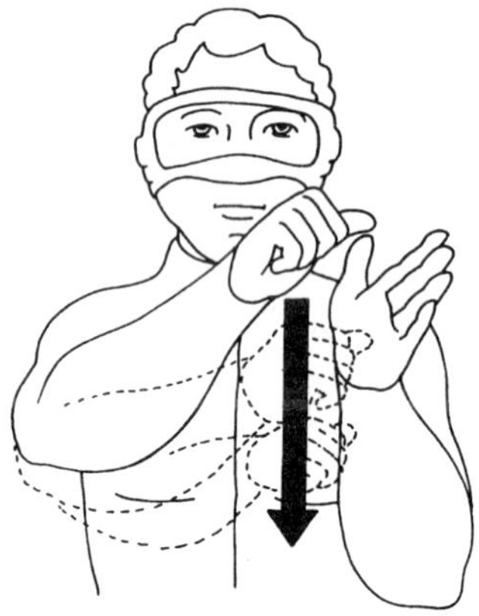

CUT / WOUND

The extended thumb of the right “A” hand is drawn across the palm of the left open hand representing the palm being cut by a knife. Then point to the area of the wound or cut.

(115)

DANGEROUS

The back of the thumb of the right “A” hand, (palm facing left) is brushed up and against the back of the left “A” hand (palm facing the diver) several times.

(116)

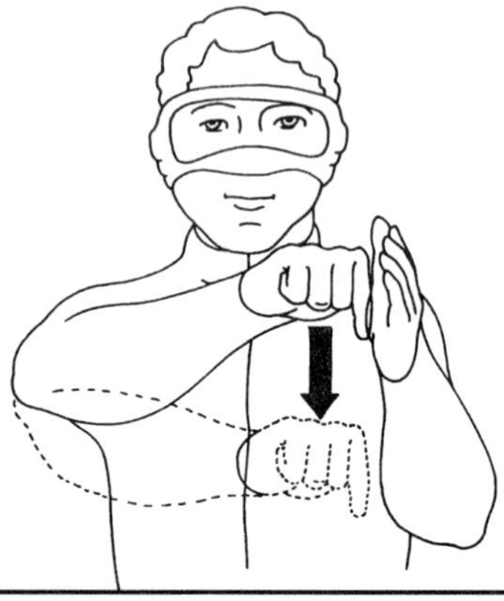

DEEP

The index finger of the right hand is pushed down the inside of and past the left open hand (palm facing right). The action of the finger moving downward indicates depth.

(117)

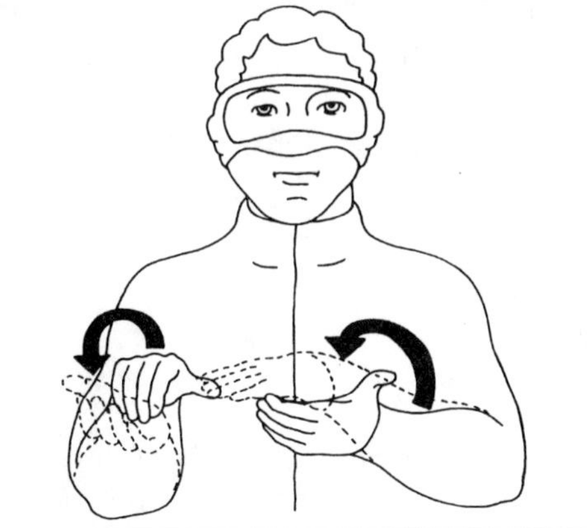

DIE

The hands are placed in front of the body (right hand, palm down; left hand, palm up) - now turn both hands over to indicate a person "turning over in his grave."

(118)

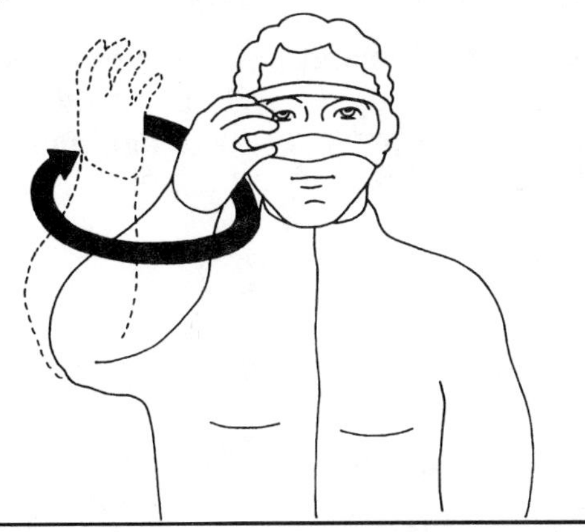

DIZZY

The right "5" hand (fingers bent, palm facing the forehead) is circled in a manner representing the spinning sensation caused by dizziness.

(119)

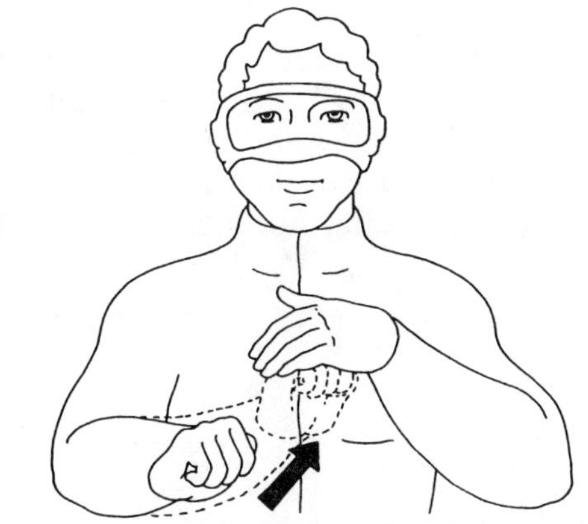

EMBOLIZE

The right fist is forcefully hit into the left palm either in the chest area or the area of the head depending on where the embolism occurs. This movement represents the air bubble (fist) affecting either the cardiovascular system or the brain.

(120)

EMERGENCY

The thumbtip of the right "L" hand is struck against the palm of the left open hand (palm facing right) two times, representing an emergency bell ringing.

(121)

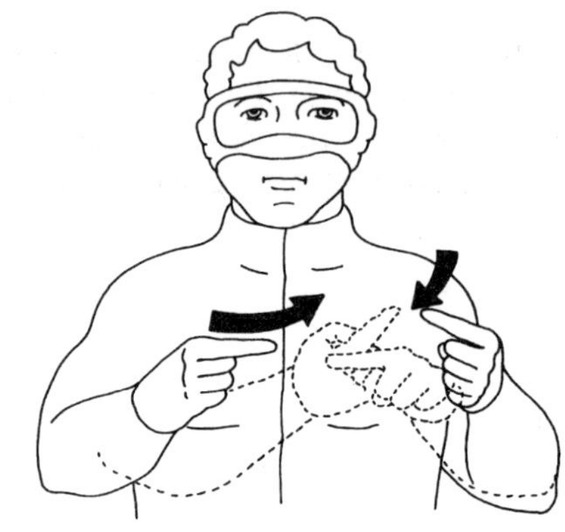

HEART ATTACK

Point to the heart with the right index finger and then jab the index fingers toward each other in a violent twisting motion in the area of the heart.

(122)

HELP

The right "A" hand (palm facing left) is lifted upward by the left open hand. This movement represents one hand being "helped" or lifted up by the other.

(123)

HURT / PAIN

The index fingers are jabbed toward each other in a twisting motion representing pain shooting through the body. *Note: it is proper to use this sign at the actual point of pain: headache, stomach ache, pain in the rear.

(124)

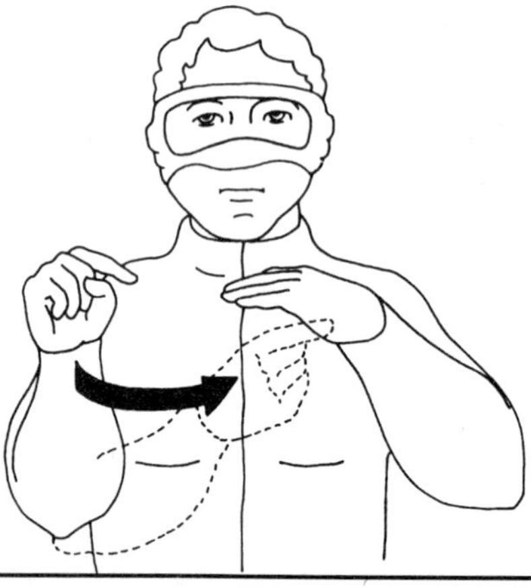

KILL

The right index finger is shoved beneath the left palm and out toward the left in a twisting movement which represents the jabbing in and twisting of a knife.

(125)

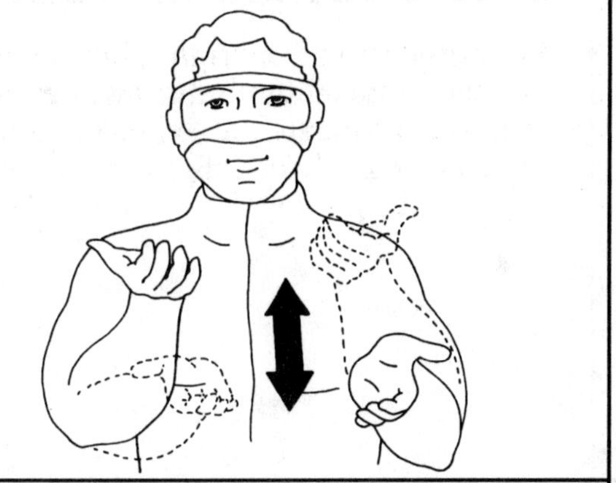

MAYBE

The open hands (palms up) are placed in front of the diver and alternately moved up and down to represent a scale and an idea or possibility being weighed.

(126)

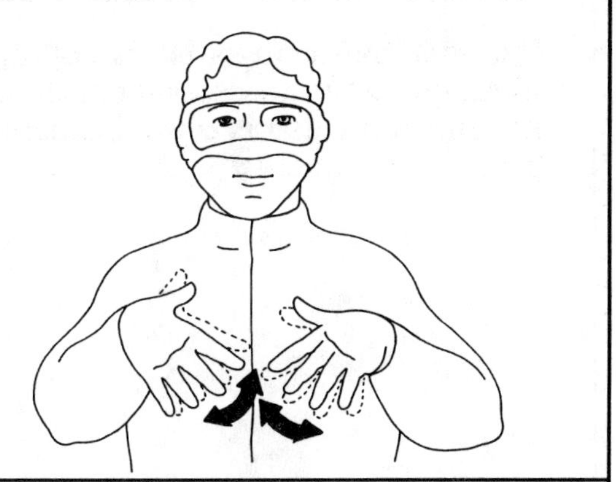

NERVOUS

The “5” hands (palms down in front of the diver) are “trembled” to indicate a nervous uncontrollable shaking.

(127)

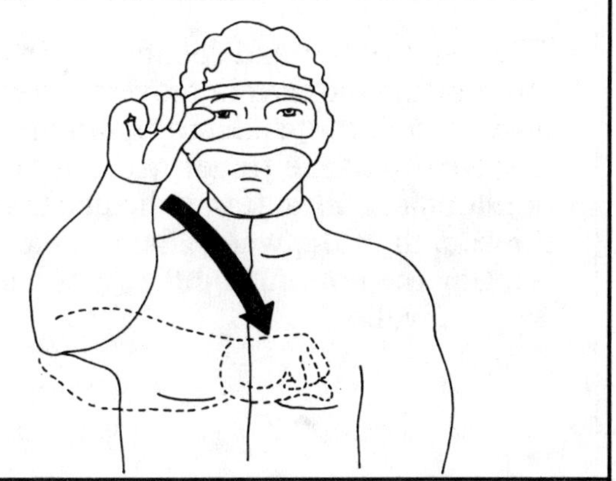

NITROGEN NARCOSIS

The sign “drunk” is used. The “A” hand (thumb extended and palm facing forward) is drawn past the lips and downward representing the town drunk trying to put a bottle to his lips and missing.

(128)

NO

The straight index and middle fingers are touched to the thumb indicating a mouth closed saying “no.”

(129)

OK / ALL RIGHT

The fingertip of the right index finger touches the right thumbtip while the remaining fingers are extended pointing upward (the palm facing left). This is the universal sign for "everything is OK!"

(130)

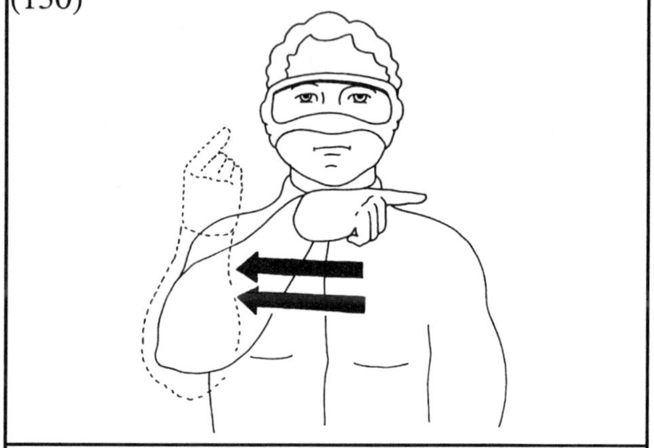

OUT OF AIR

The right index finger is drawn from left to right two times across the neck indicating air supply that is cut off or depleted.

(131)

SHALLOW

The right open hand (palm and fingers pointing to the left) is moved across the top of the left open hand (palm up) from the wrist past the fingertips. The hands do not touch! This action indicates the shallow depth of the water; the left hand representing the bottom and the right hand the water level.

(132)

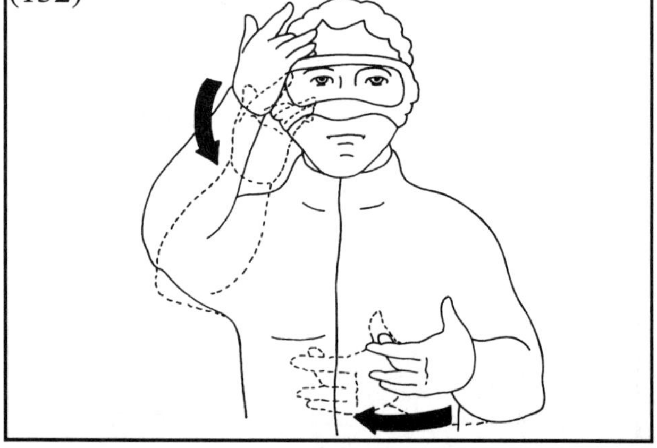

SICK

The middle finger of the right "5" hand is touched to the forehead while the middle finger of the left "5" hand is touched to the stomach to indicate discomfort or illness in these areas.

(133)

STOP

The right open hand (palm facing left) is abruptly brought down into the left open hand (palm up).

(134)

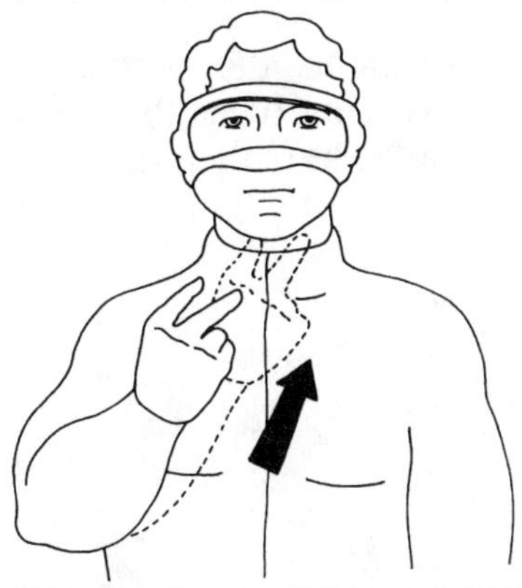

STUCK

The index and middle fingertips of the right “V” hand (palm down) are touched to the front of the neck. This sign usually refers to being trapped or caught in an untenable position.

(135)

TEMPERATURE

The right index finger (pointing left) is moved up and down the outside of the left index finger (pointing up) several times to represent the mercury in a thermometer moving up and down.

(136)

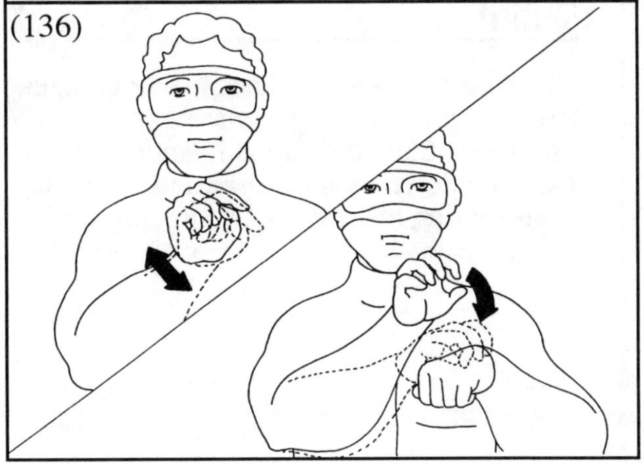

THAT BITES

Point to the creature in question; then the “5” hand (fingers bent representing a mouth) “bites” the back of the left hand, pinching the skin.

(137)

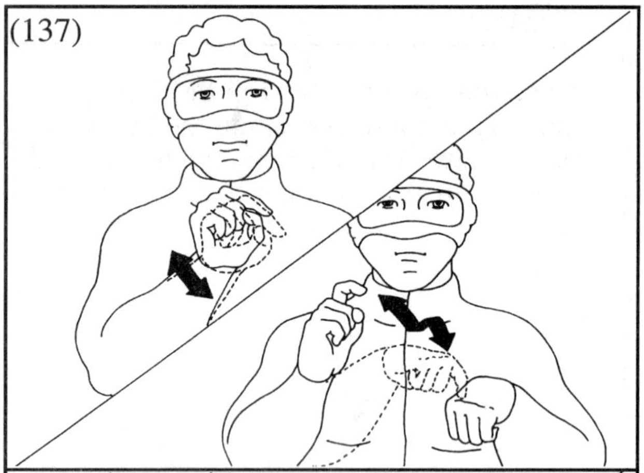

THAT STINGS

Point to the creature in question; then with the index finger (stinger) of the closed right hand jab the back of the left hand several times representing the stinging action of an underwater animal.

(138)

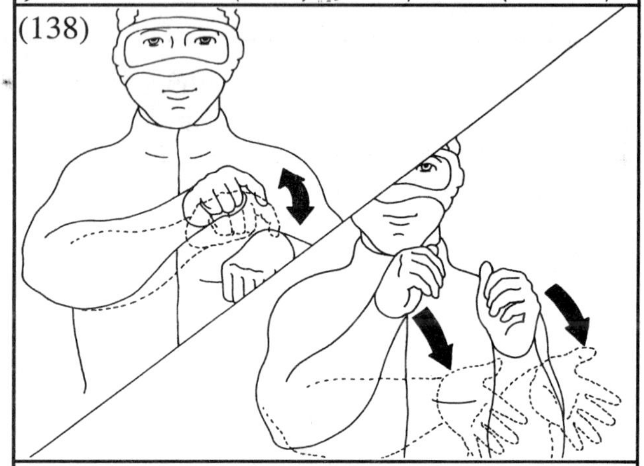

TIME REMAINING

The diver touches the top of his left wrist or his watch with his right index finger (indicating "time") then the "5" hands (palms facing, fingers pointing up) are abruptly lowered to a fingers-pointing forward position.

(139)

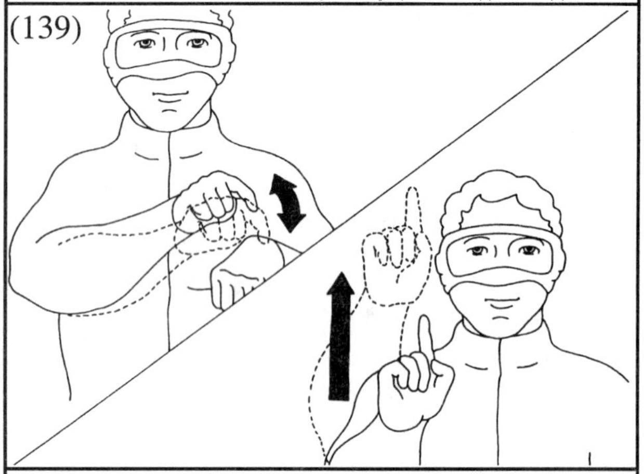

TIME TO GO UP

The diver touches the top of his left wrist or his watch with his right index finger (representing the "time") and then points toward the surface.

(140)

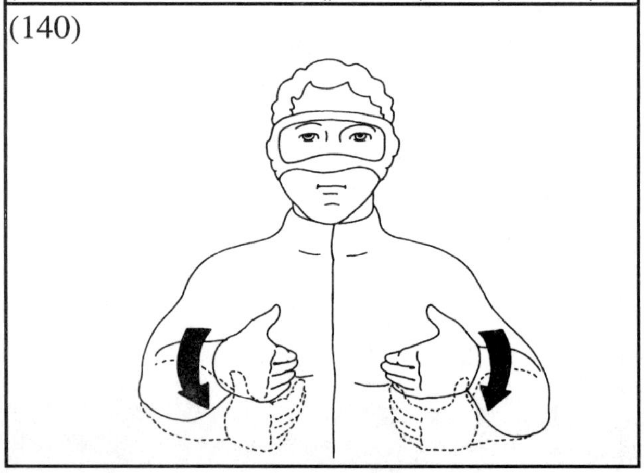

TIRED

The fingertips of both hands (palms facing the body) are touched to the upper abdomen; then, the cupped hands are dropped downward to a palm up position representing the tired body sagging down.

(141)

VOMIT

The index finger of the right "5" hand (palm facing left) is touched to the lips and the hand is arched out and downward representing the vomitus coming from the mouth.

Chapter 6

DIVING FOR A LIVING

(The Commercial Diver)

The signs in this chapter are designed specifically for use by the commercial or salvage diver. However, the signs presented will certainly be of value to the scuba diver as well. It is important that the underwater communicator be skilled in all aspects of the below-the-surface communicative process.

(142)

ABOVE

The right open hand (palm down) is held above the left open hand (palm down) and circled in a clockwise direction. This movement represents the idea that something is moving above a stationary object.

(143)

BEHIND

The right "A" hand (palm facing left, and in front of the left "A" hand which is facing right) is moved back and "behind" the left "A" hand.

(144)

BELOW / BENEATH

The right open hand (palm down) under the left open hand (palm down) is circled in a clockwise direction. This movement represents the idea that something is moving beneath a stationary object.

(145)

BETWEEN

The right open hand (palm facing left shoulder) is placed between the second and third fingers of the left open hand (palm facing right) and moved up and down several times. This sign represents an object that is placed between two other objects.

(146)

BOARD / WOOD

The right open hand (palm facing left) and little finger edge touching the back of the left hand (palm down) is "sawed" back and forth across the back of the left hand. This action represents wood being sawed.

(147)

BREAK

Both "S" hands are held side by side (palms down) in front of the diver as if holding an imaginary "stick." The stick is then "broken" so that the palms of the "S" hands are facing each other.

(148)

BREAKDOWN (as in engine failure)

The fingertips of the "5" hands (palms down) are intertwined and then abruptly dropped downward indicating a motor or building collapsing.

(149)

BRIDGE

The left arm (palm facing down) is touched at the wrist, forearm and elbow from beneath by the fingertips of the right "V" hand. The right "V" hand, as it touches the left arm (bridge) represents support pillars.

(150)

BUILD

The palm of one hand is placed on the top of the other hand in an alternating, ascending manner indicating one brick being placed on top of another brick to construct a building.

(151)

BUSY

The wrist of the right "B" hand (facing forward) is moved back and forth several times across the inside of the wrist of the left "S" hand (palm down). This is basically the same sign as "work," except the "B" handshape is used.

(152)

CHANGE

The "X" hands (palms facing) touch and stay together; the hands are then twisted so that the index finger of the right "X" hand is on the outside (away from the body) and the index finger of the left "X" hand is on the inside (near the body). This movement represents the changing positions of the "X" hands.

(153)

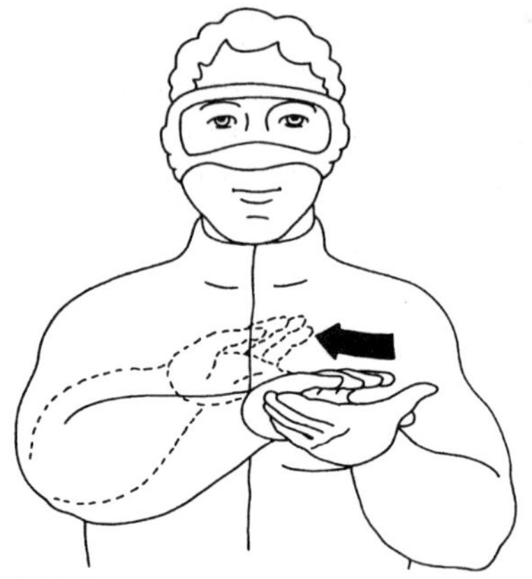

CLEAN

The right open hand (palm down) is slid across the left open hand (palm up) from the base of the thumb past the fingertips several times. The sign indicates crumbs being wiped from a table in order to "clean" it.

(154)

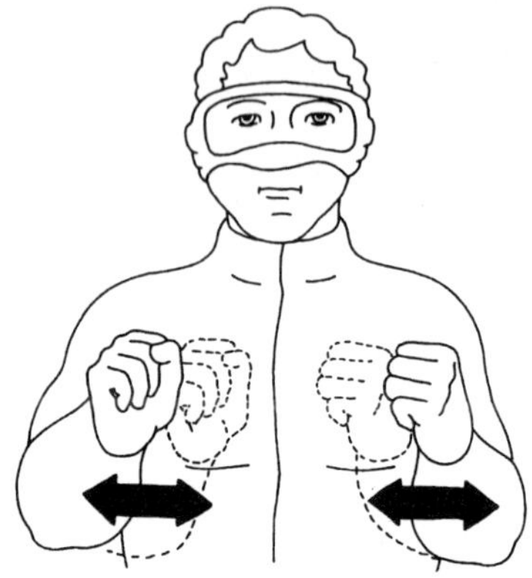

COLD

The "S" hands are held at the side of the diver (palms facing at chest level) and simultaneously moved back and forth toward one another several times to indicate the act of shivering.

(155)

COMBINE / JOIN / CONNECT

Both "5" hands are brought together allowing the index finger and thumb of one hand to interlock with the index finger and thumb of the other hand. This movement represents the linking together of a chain.

(156)

COMPARE

The right open hand (palm facing left, fingers extended pointing up) is turned so that the palm faces the diver. Alternatively, the left open hand (palm facing right, fingers extended pointing up) is turned so that the palm faces the diver. This action is repeated in an alternating fashion several times representing the palms being examined on a comparative basis.

(157)

DIFFERENT

The index fingers are crossed in front of the diver (palms out) and then pulled apart pointing outward.

(158)

DIRTY

The back of the right “5” hand (palm down and fingers pointing left) is touched beneath the chin and the fingers are then wiggled up and down.

(159)

DIVIDE / SEPARATE

The fingers of both open hands are placed back to back with the fingertips facing the diver and then the hands are drawn apart. This movement represents something being pulled apart or “separated.”

(160)

DON’T WANT

The “5” hands (palms up and fingers slightly bent) are placed in front of the diver and then turned over to a palms down position.

(161)

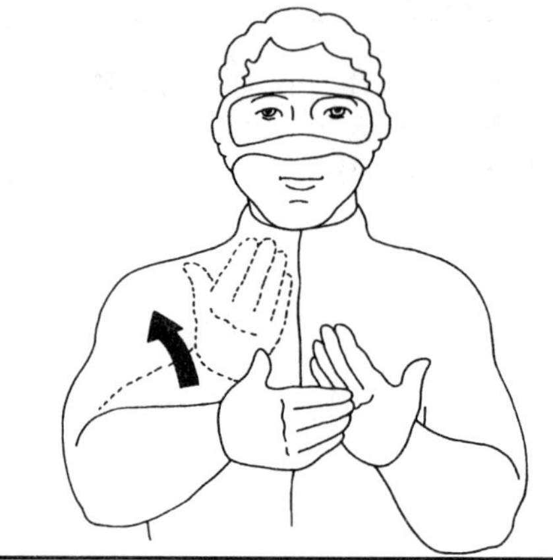

EASY

The fingertips of the right open hand (palm facing left shoulder) are brushed upward against the back of the fingertips of the left open hand (palm facing the diver) several times, indicating the ease of movement of the fingertips as they are drawn upward.

(162)

EMPTY / BARE

The middle finger of the right "5" hand is drawn across the back of the left hand (palm down) from wrist to fingertips. This sign represents the relative absence of hair (bareness) on the back of the hand.

(163)

ENGINE / MOTOR / MACHINE

The fingers of the "5" hands (palms facing the diver) are intertwined (fingertips towards the body) and then the hands are moved up and down several times representing the gears of an engine meshing together.

(164)

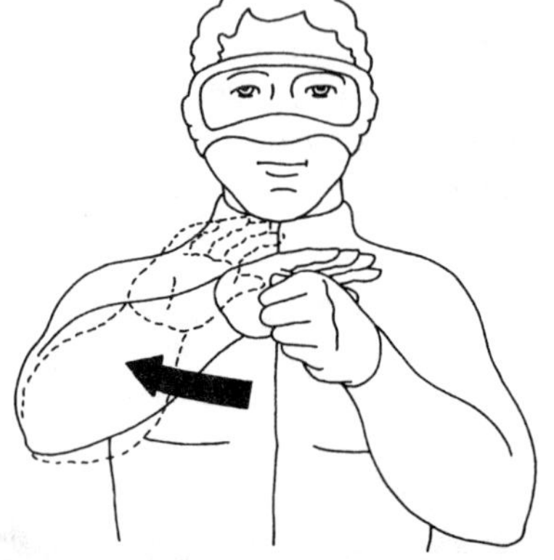

ENOUGH

The right open hand (palm down) is brushed twice (away from the body) across the top of the left "S" hand (palm facing left).

(165)

EXPLODE

With the palms facing out, the "S" hands are touched in front of the diver and then "exploded" upward and to the sides ending in a "5" hand position.

(166)

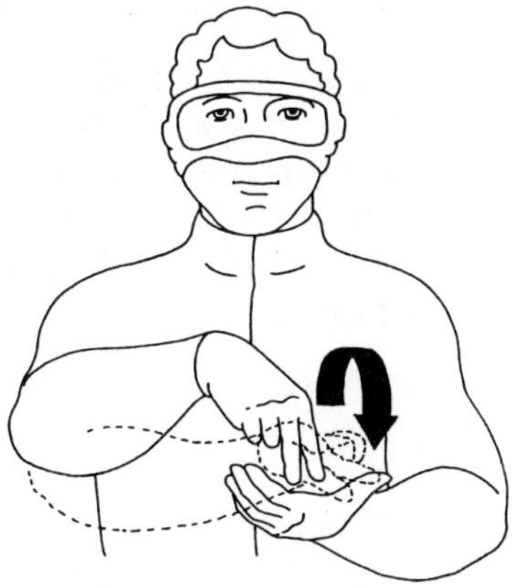

FALL

The index and middle fingers of the right "V" hand are touched to the left palm (facing up) and then the "V" falls into the left hand. This movement represents a person standing up and then falling onto his back.

(167)

FIRE

With the palms of the "5" hands (at chest level) facing the diver, the hands are moved up and down several times and simultaneously the fingers are wiggled representing flames leaping upward.

(168)

FULL / FILL

The right open hand (palm down) is drawn toward the body across the top of the left "S" hand (palm facing left). This sign is representative of a cup or similar container that is "full" to the brim.

(169)

GASOLINE

The extended thumb of the right closed hand is stuck into the top of the left "S" hand (palm facing right) twice. This movement represents the gasoline nozzle being stuck into a can in order to pour the gasoline.

(170)

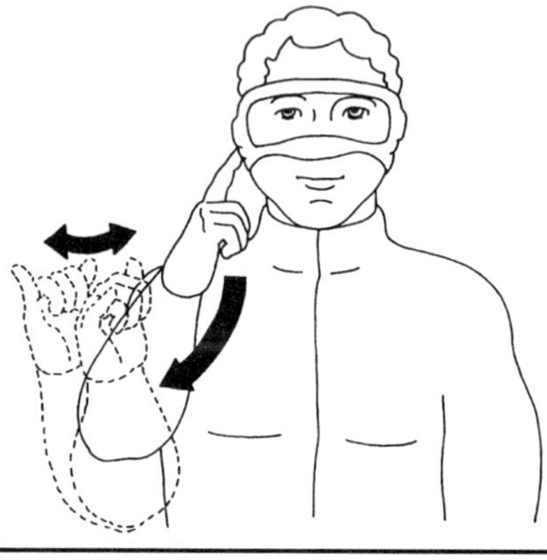

GOLD

The index finger of the right hand is touched to the right ear lobe and then the hand is twisted out (palm facing forward) to a "Y" position. This sign represents "gold or yellow" earring. *This sign also represents California.

(171)

HARD (as in "not soft")

The middle finger of the right bent "H" hand is struck on the back of the left "S" hand. This movement represents the hardness of the back of the hand.

(172)

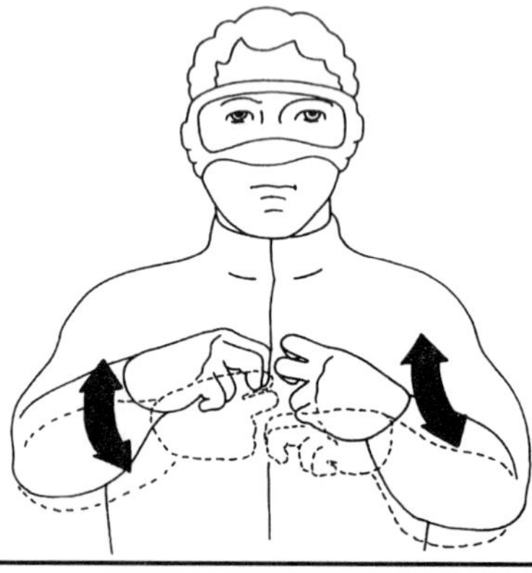

HARD / DIFFICULT

The knuckles of the bent "V"s (palms facing the diver) are struck together several times in an alternating up and down movement. This movement represents an attempt to link the gears (bent "V"s) together but it is difficult.

(173)

HOT

The index and middle fingertips of the right "C" hand are touched to the lips or regulator and then quickly the hand is drawn away and turned downward indicating a "hot" cup of coffee being brought to the mouth and quickly removed.

(174)

HOUR

The right closed hand (index finger extended) is placed palm down in the center of the left open hand (palm facing right). Then the right wrist is rotated so that the right index finger describes a clockwise circle in the center of the left open hand. This movement represents an hour hand of a clock making a full cycle or one "hour".

(175)

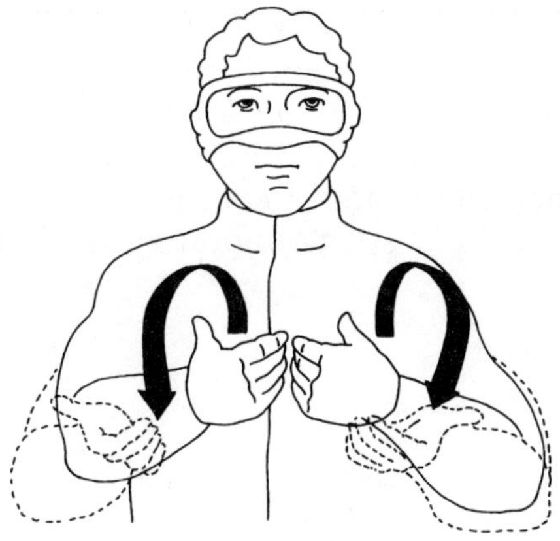

HOW

The fingertips of the curved open hands (palms facing the diver) are placed back to back and then the hands are rolled forward to a palm up position.

(176)

IMPORTANT

The "F" hands are moved up and in front of the diver until the index fingertips and thumbtips touch (right palm facing left; left palm facing right).

(177)

INCH

The little finger of the right "I" hand is moved back and forth across the index finger (pointing right) of the left hand representing a span of approximately one inch.

(178)

CENTIMETER

This is basically the same movement as "inch" except the "C" hand is moved back and forth across the index finger representing a span of approximately one centimeter.

(179)

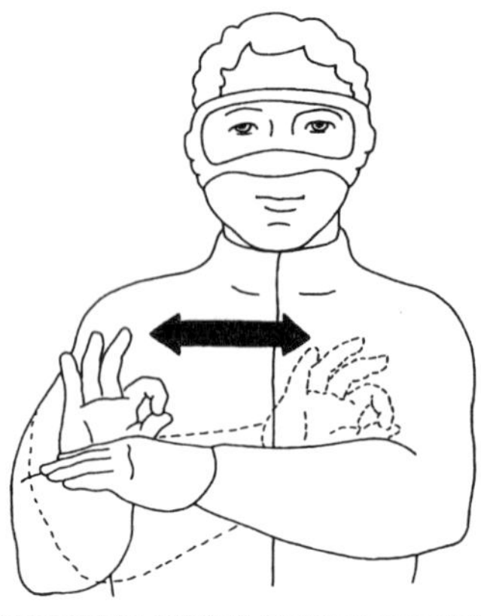

FOOT

The right "F" hand (palm facing forward) is moved back and forth across the inside of the left arm (palm facing down) from fingertips to mid forearm to fingertips representing a span of approximately one foot.

(180)

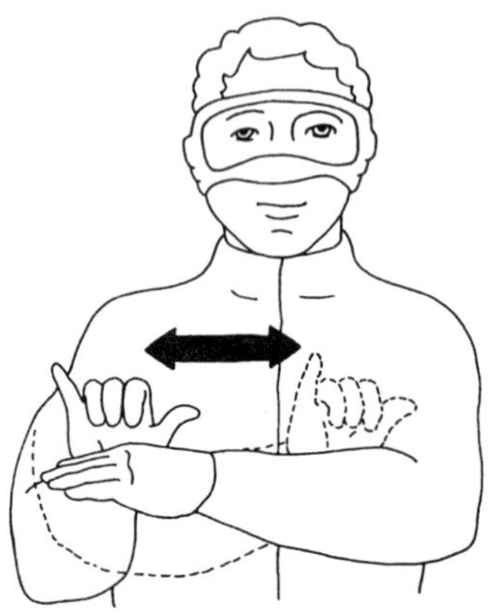

YARD

The right "Y" hand (palm facing forward) is moved back and forth across the inside of the left arm (palm facing down from fingertips to elbow to fingertips representing a span of approximately one yard.

(181)

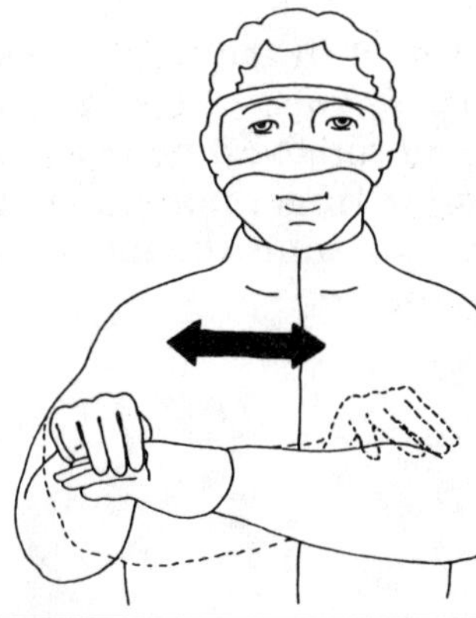

METER

The same movement as employed in "yard" except the right "M" hand is used representing a span of approximately one meter.

(182)

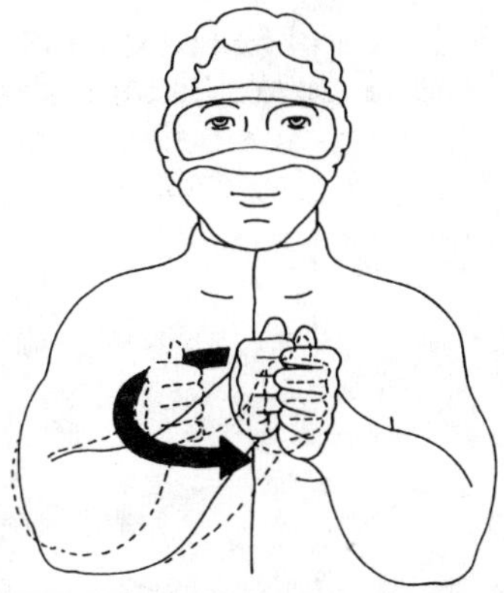

IN FRONT OF / PASS

The right "A" hand (palm facing left and behind the left "A" hand which is facing right) is moved past and "in front of" the left "A" hand.

(183)

JEWELRY / JEWEL

The middle fingertip of the right "5" hand is touched to the back of the ring finger of the left "5" hand (palm down) and then moved upward and away in a shaking motion. This movement signifies the glittering of a precious stone.

(184)

LARGE / BIG

Both "L" hands (palms facing and index fingers pointing out) are drawn apart indicating the "bigness" of the object described. The size of the sign is indicative of just how large the object is.

(185)

LATER / AFTER AWHILE

The thumbtip of the right "L" hand is touched to the center of the left palm (facing right) and rotated downward so that the right index finger points down. This movement represents the movement of a hand on a clock.

(186)

LEFT (direction)

The left "L" hand (palm out) is placed in front of the diver and then drawn to the "left."

(187)

LONG

The right "L" hand (palm down) is drawn up the length of the left arm (palm down) from fingertips to shoulder indicating length involved in measurement.

(188)

MEASURE

The "Y" hands are placed in front of the diver (palms down) then the thumbtips are touched together several times indicating approximately 12" from little finger to little finger.

(189)

METAL

The fingertips of the right "M" hand are tapped twice on the back of the left "S" hand. This movement represents the hardness of metal.

(190)

MINUTE

The right closed hand (index finger extended) is placed palm down in the center of the left open hand (palm facing right). The index finger is then moved toward the little finger of the left open hand. This sign represents the movement of the minute hand on a clock.

(191)

MONEY

The right open hand (palm up) is tapped twice into the left open hand (palm up). This movement represents the act of placing dollar bills into an open hand.

(192)

MORE

The thumb and fingertips of the "and" hands (palms facing) are brought together in front of the diver. The sign represents a quantity being added to another quantity to create "more" quantities.

(193)

MOVE / PUT

Both hands are placed in front of the diver (palms down and thumbtips and fingertips of index and second fingers touching). Now the hands are lifted and moved to one side as if an object is being picked up and put in another place.

(194)

MUST / NEED / NECESSARY

The right "X" hand (palm facing down) is abruptly moved downward.

(195)

NOT YET

The right hand (palm facing back and fingers pointing down) is placed near the right side of the body and waived back and forth several times.

(196)

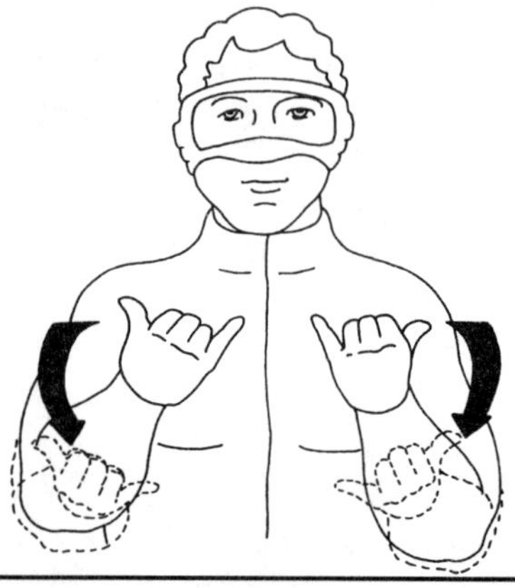

NOW

Both "Y" hands (palms up in front of the diver) are dropped slightly downward indicating time that is at hand (now).

(197)

OIL / GREASE

The open left hand (palm facing the diver; fingers pointing right) is grasped at its lower edge by the right index finger and thumb. The index finger and thumb are drawn downward several times representing the dripping of oil.

(198)

PRESSURE

The thumbtip of the right "A" hand is pressed into the palm of the left open hand and twisted twice in a clockwise direction.

(199)

RIGHT / CORRECT

The right closed hand (palm facing left, index finger pointing forward) is struck once on the top of the left closed hand (palm facing right, index finger pointing forward).

(200)

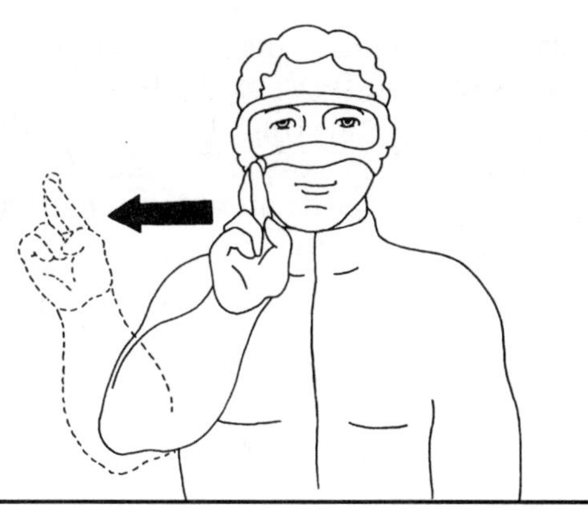

RIGHT (direction)

The right "R" hand (palm out) is placed in front of the diver and then drawn to the right.

(201)

SAME

The sides of the index fingers (pointing away from the body; palms down) are struck together twice.

(202)

SHARP

The tip of the middle finger of the right "5" hand is touched to the underside of the left open hand (palm facing right) and then quickly moved away so that the right palm faces downward. This movement represents touching the sharp edge of a knife and then quickly withdrawing the finger.

(203)

SHORE / BEACH

The fingers of the right "5" hand (palm down) are wiggled across and back down the left open hand (palm up) several times. This movement represents wave action as the surf rolls up the beach and then abates.

(204)

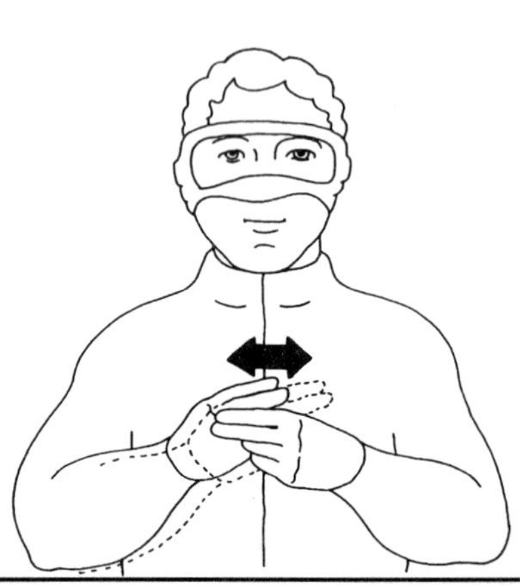

SHORT / SOON

The middle finger of the right "H" hand is rubbed back and forth several times (fingertip to knuckle to fingertip) across the index finger of the left "H" hand. The shortness of the movement is indicative of the length described.

(205)

SILVER

The index finger of the right hand is touched to the right ear lobe and then the hand is twisted out (palm facing forward) to an "S" position. This sign represents a "silver" earring.

(206)

SMALL

Both open hands (palms facing and fingers extended) are drawn together until they almost touch. The size of the sign is indicative of just how small the object is.

(207)

SOFT

Both "5" hands (palms facing up) are simultaneously drawn downward (thumbs touching fingertips). This movement is repeated twice and represents feeling a soft piece of cotton between the thumb and fingertips.

(208)

START / BEGIN

The index finger of the right closed hand is stuck between the index finger and second finger of the open hand (palm facing left) and twisted clockwise. This movement represents a key being turned to "start" a car.

(209)

TRADE / EXCHANGE / REPLACE

The left modified "X" hand position (thumbtip touching index fingertip; palm facing right) is placed in front of the right modified "X" hand position (thumbtip touching index fingertip; palm facing left); the hands then "trade places" with the right hand going under and in front of the left as the left moves over and behind the right hand.

(210)

WANT

The "5" hands (palms up) are placed in front of the diver and then drawn toward the body as the fingers are flexed. This action represents the pulling of a "wanted" object toward the diver.

(211)

WELD

The index finger of the right hand (thumb extended and palm facing left shoulder) is moved back and forth across the palm of the left hand. This movement represents the action of a blowtorch in the welding process.

(212)

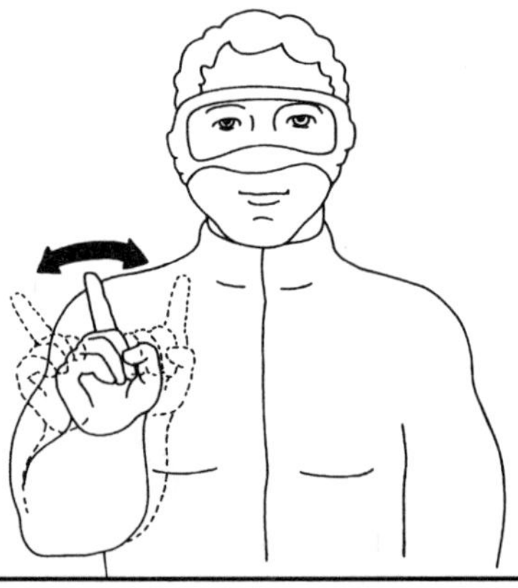

WHERE

The right index finger is held pointing up (palm out) in front of the diver and the hand is shaken from side to side several times.

(213)

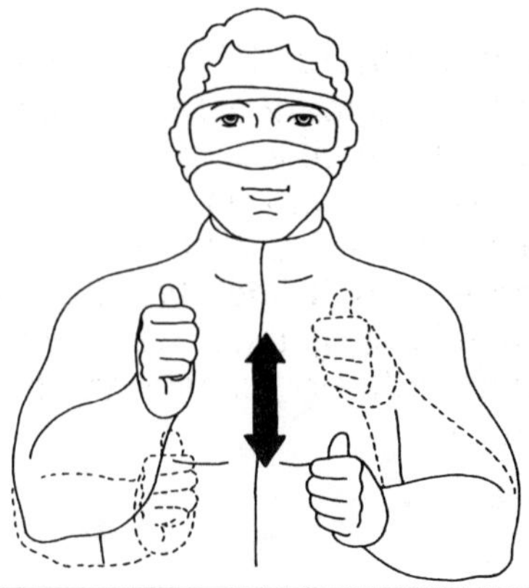

WHICH

The "A" hands (palms facing) are held in front of the diver and alternately moved up and down several times indicating a "choice" to be made as to "which" one.

(214)

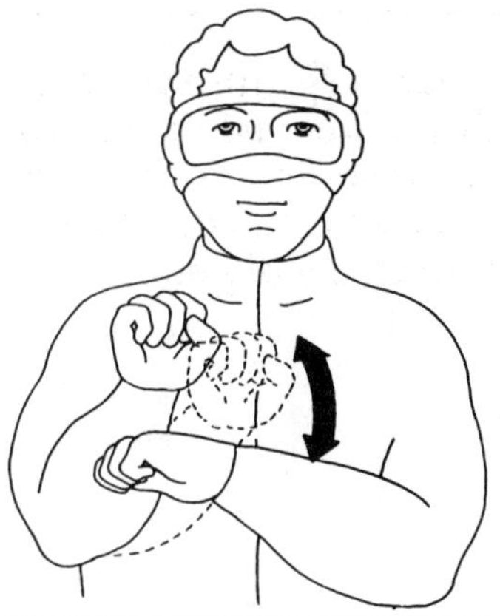

WORK

The wrist of the right "S" hand (palm facing forward) is struck twice against the inside of the wrist of the left "S" hand (palm down).

(215)

WRONG / MISTAKE

The "Y" hand (palm facing the diver) is touched to the chin.

Chapter 7

MISCELLANEOUS SIGNS

The signs selected for this section of the text are by no means merely a hodgepodge of unrelated symbols. Rather, they encompass a carefully selected lexicon of signs related to and applicable to concepts which have been presented in previous chapters.

With a mastery of the symbols in "Miscellaneous Signs" and a working knowledge of the signs already covered, the scuba or commercial diver is well on the way to becoming, in the fullest sense of the word, an underwater communicator.

(216)

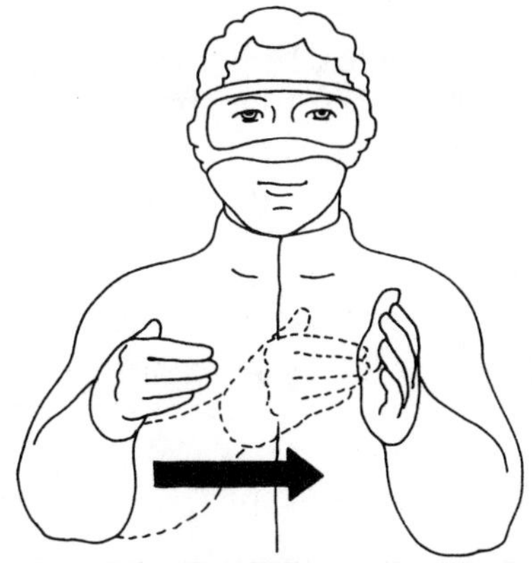

AGAIN

The fingertips of the bent right open hand (palm left) are touched once to the middle of the left open hand (palm right).

(217)

ALL

The right "open" hand is circled once in a clockwise manner around the left open hand (palm up) and then placed in the left open hand (both palms up). The movement of the right hand represents collecting all of a substance and placing it in the left hand.

(218)

BAD

The fingertips of the right "open" hand are touched to the regulator/lips and then the hand is lowered to a palm down position in front of the diver.

(219)

BEAUTIFUL / PRETTY

The "5" hand (palm facing diver's face) is circled in a counterclockwise manner ending with the thumbtips and fingertips touching to the right of the face.

(220)

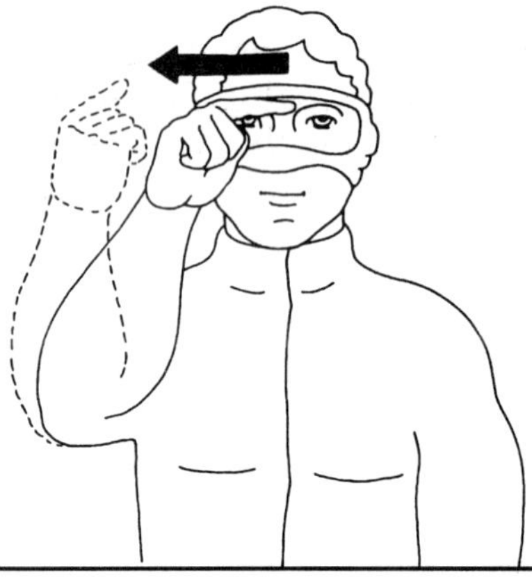

BLACK

The right index finger (palm facing left) is drawn from left to right across the forehead above the right eyebrow indicating the "black" color of the eyebrow.

(221)

BLUE

The right "B" hand (palm facing left; fingers pointing up) is quickly switched to a palm downward position. This sequence is repeated twice.

(222)

BRIGHT / CLEAR

The fingertips and thumbtips of both hands touch (palms facing). Then both hands are opened into "5" handshapes (palms out) as they simultaneously arc to the sides. This movement indicates light rays spreading out in dazzling brightness.

(223)

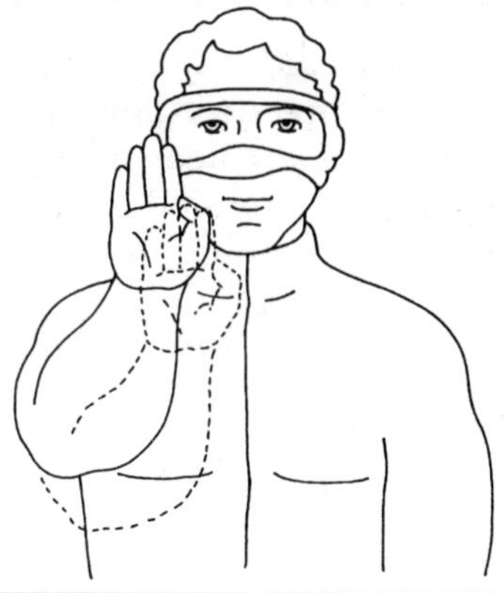

BROWN

The right "B" hand (palm forward) is drawn twice down the right cheek. This movement indicates the "brown" color of the skin.

(224)

CAMERA

The small imaginary camera is held in front of the face (palms facing in modified "C" positions) and the right index finger is moved up and down as if "taking a picture."

(225)

CAN

Both "A" hands (palms down) are placed side by side in front of the diver and then simultaneously lowered approximately six inches.

(226)

CAN'T

The right index finger (palm down) strikes the tip of the left index finger (palm down) as it moves downward. This movement indicates the left index "can't" support the right index.

(227)

COCONUT

The thumbtips and fingertips of the "5" hands are touched together as if holding a coconut and shaken near the ear. This sign represents the act of shaking a coconut to hear the milk splashing inside.

(228)

COLOR

The fingertips of the right "5" hand (palm facing diver) are wiggled in front of the mouth or regulator.

(229)

COME

The index fingers are rotated around each other as the hands are drawn toward the diver's body. This movement represents an object/person coming toward the diver.

(230)

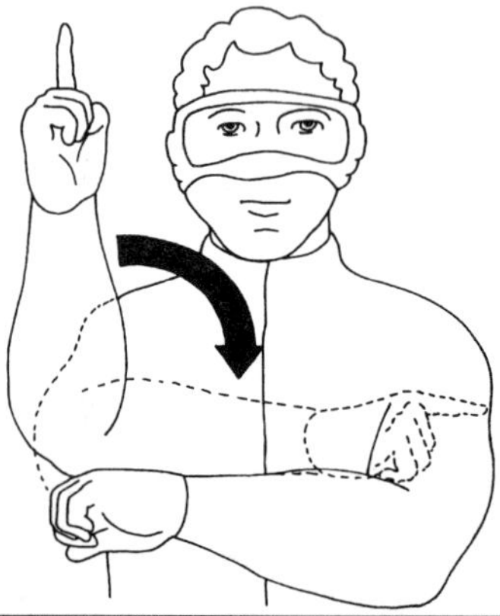

DAY

The right elbow is placed on top of the left hand (palm down), right index finger pointing up and right palm facing left. The right hand is then lowered to the left elbow. This movement represents the sun's path across the sky.

(231)

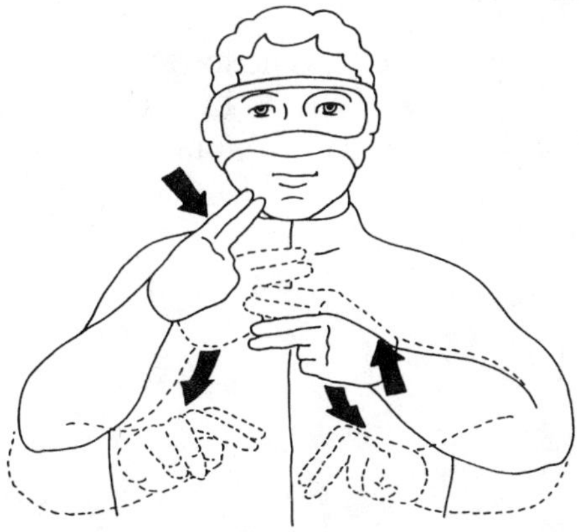

EGG

The middle finger of the right "H" hand (palm facing left shoulder) is struck on top of the index finger of the left "H" hand (palm facing right shoulder); the "H" hands now drop to either side (palms down). This action resembles the cracking of an egg preparatory to cooking it.

(232)

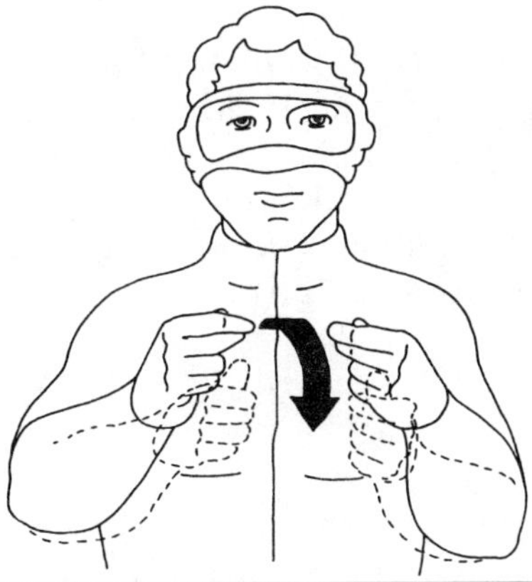

FAST

The thumbtips are touched to the index fingertips and then snapped out once or twice representing the idea of snapping the fingers to the beat of "fast" music.

(233)

FIND / DISCOVER

The "5" hand (palm down) is placed in front of the diver and then raised upward as the thumbtip and index fingertip simultaneously touch and lift up representing something being picked up which has been lost.

(234)

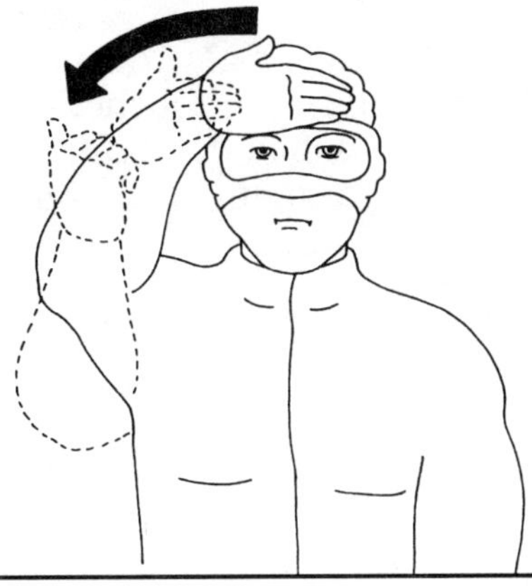

FORGET

The right open hand is wiped across the forehead ending in an "A" position (palm facing diver and right of the head). This sign represents knowledge that has been erased or wiped from the mind.

(235)

FUN

Touch the tip of the nose/regulator with the fingertips of the right "H" hand; then place the right "H" fingers (palm down) on top of the left "H" fingers (palm down).

(236)

GIVE

Both hands (thumbtip and fingertips touching) are placed side by side, palms up in front of the diver. The hands are then pushed forward changing into open hand positions (palms up). This sign represents the offering of a "gift."

(237)

GO

The index fingers are rotated around each other as the hands are moved away from the diver's body. This movement represents an object/person going away from the diver.

(238)

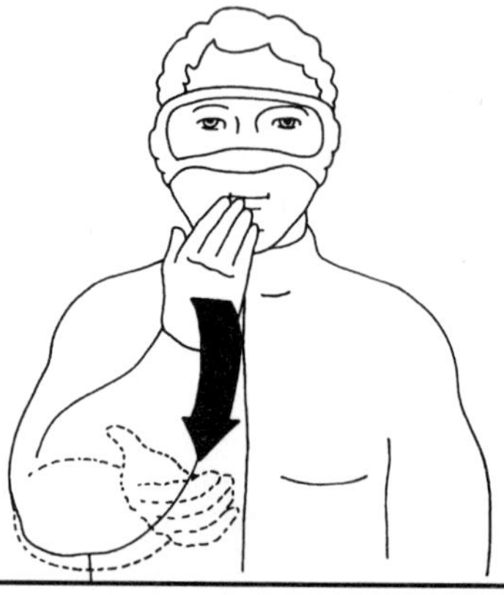

GOOD

The fingertips of the right "open" hand (palms facing diver) are touched to the regulator/lips and then the right open hand is lowered to a palm up position in front of the diver.

(239)

GREEN

The right "G" hand (palm facing left) is shaken several times in front of the diver.

(240)

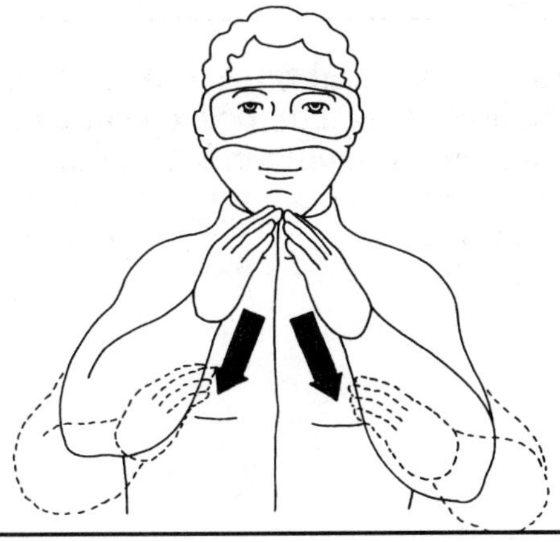

HOUSE

The middle fingertips of the open hands touch forming the angle of the "roof of a house;" then the hands are drawn downward indicating the slant of the roof.

(241)

HURRY

The right "H" hand (palm facing left) is rapidly shaken up and down in front of the diver.

(242)

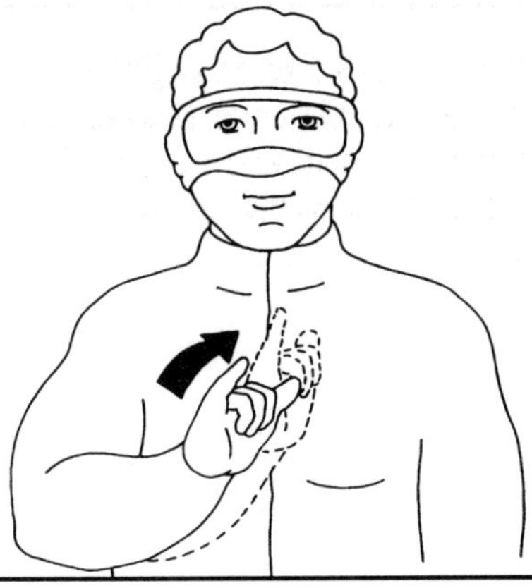

I

The thumb of the "I" hand is touched to the chest.

(243)

KNOW

The fingertips of the right "open" hand are touched to the right forehead.

(244)

LIKE

The thumbtip and fingertips of the "5" hand are touched to the chest and then drawn forward, thumbtip and index fingertip touching.

(245)

LOOK

The fingertips of the "V" hand (palm facing diver) are touched to the face plate of the mask; then the "V" hand is turned palm out with the fingertips pointing forward representing tiny eyes on the "V" fingertips "looking" around.

(246)

LOVE

The crossed "S" hands (palms facing the diver) are placed over the heart.

(247)

MAN / MALE

The brim of an imaginary hat is grasped between the thumb and fingertips and tipped forward. The approximate height of the male may then be indicated by placing the open hand (palm down) at a level corresponding to his height.

(248)

ME

The index finger is touched to the chest.

(249)

MINE / MY

The open palm is touched to the chest.

(250)

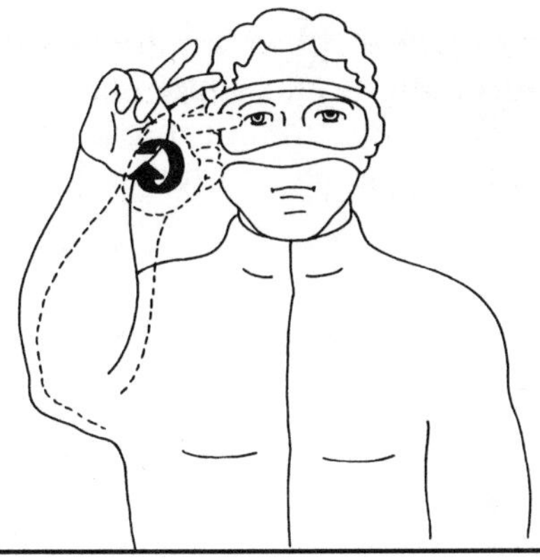

MISUNDERSTAND

The index finger of the right "V" hand (palm out) is touched to the forehead, then the middle finger of the right "V" hand (palm in) is touched to the forehead. This sign represents the thought being reversed or misunderstood.

(251)

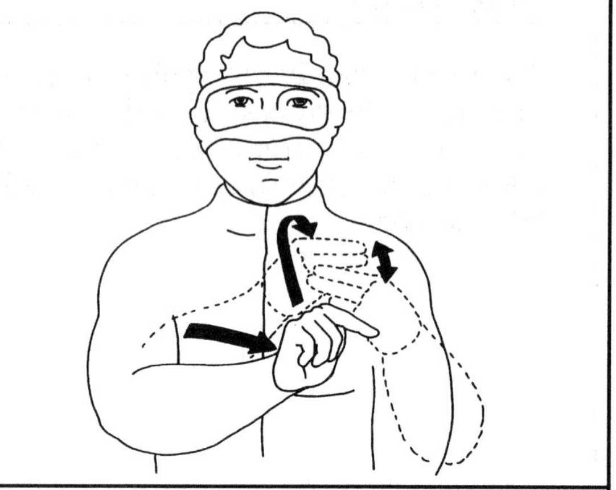

NAME OF .....

The object or creature in question is pointed to and then the middle finger of the right "H" hand (palm facing left shoulder) is tapped twice on top of the index finger of the left "H" hand (palm facing right shoulder.)

(252)

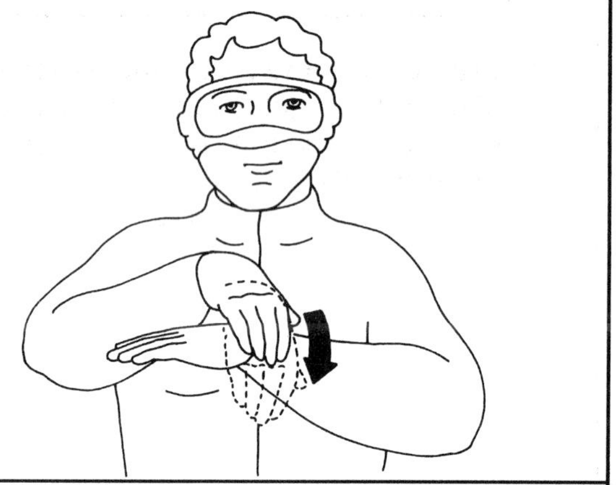

NIGHT

The right open hand (palm down) is draped across the back of the left wrist (left palm down). This movement represents the sun as having gone down.

(253)

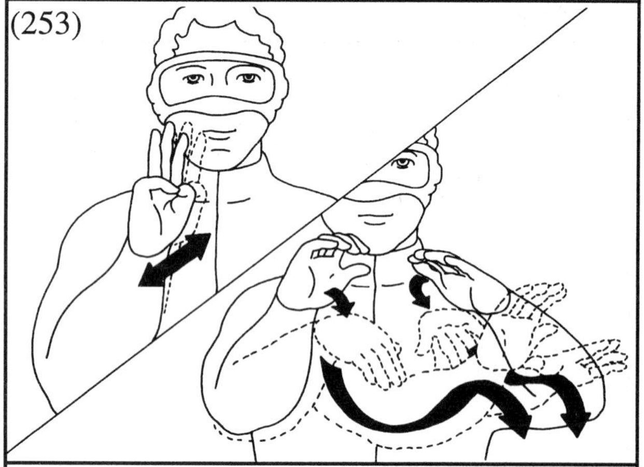

OCEAN / SEA / GULF (dual sign)

a. Water
The index fingertip of the right "W" hand (palm facing left) is touched to the lower lip or regulator.
b. Waves
Place the right "5" hand behind the left "5" hand (both palms down); move the hands forward in an up-and-down wavy motion indicating the action of swells.

(254)

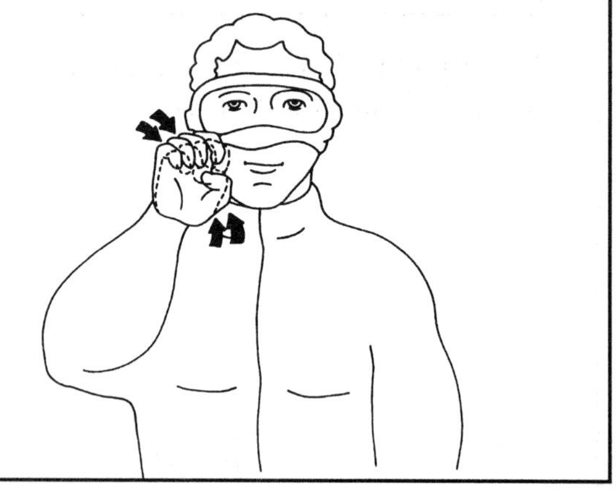

ORANGE

The "S" hand is placed at the side of the mouth/regulator and squeezed several times. This movement represents an orange being squeezed for its juice.

(255)

OUR / OURS

The right "open hand" (palm facing left) is touched to the right side of the chest and then moved in a semi-circular fashion allowing the little finger side to touch the left shoulder.

(256)

PICTURE

The right "C" hand (palm facing left) is touched to the right side of the face and then placed in the left open palm (palm facing forward) indicating a picture of the face being placed on developing paper.

(257)

PINK

The middle finger of the right "P" hand (palm facing diver) is drawn twice down the lips or regulator indicating the "pink" color of the lips.

(258)

PURPLE

The right "P" hand (palm down) is shaken twice in front of the diver.

(259)

RED

The right index finger (palm facing diver) is drawn twice down the lips or regulator indicating the "red" color of the lips.

(260)

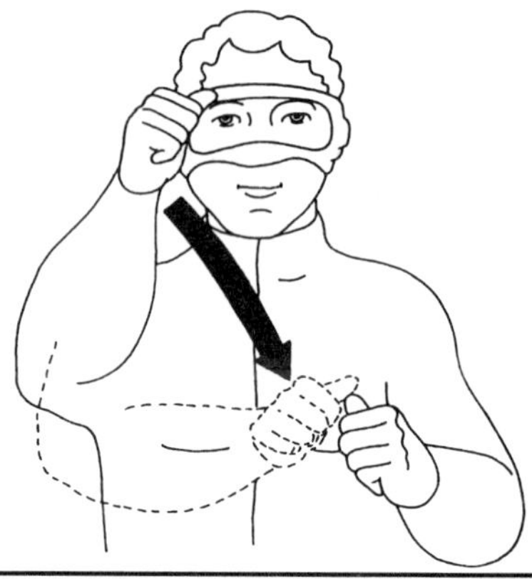

REMEMBER

The thumbtip of the right "A" is touched to the forehead and then placed on the thumbnail of the left "A" hand. This sign represents placing knowledge on your thumbnail so it will not be forgotten.

(261)

SEE / SAW

The fingertips of the "V" hand (palm facing diver) are touched to the face plate of the mask and then moved forward. This sign represents tiny eyes on the "V" fingertips which "see."

(262)

SIT

The right "V" fingers (palms down) are placed across the left "V" fingers (palm down). The "V" fingers represent a person's legs in a sitting posture.

(263)

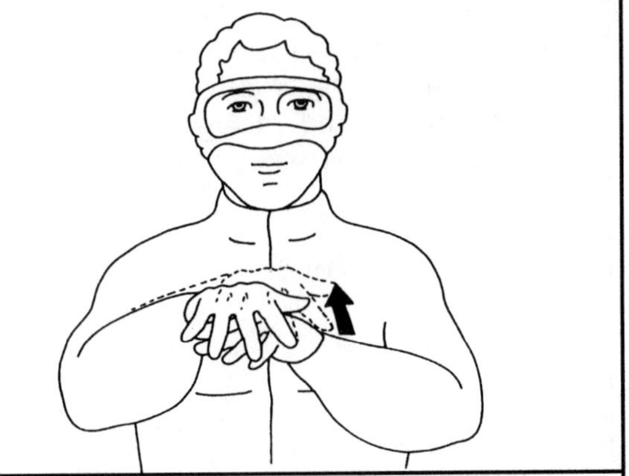

SLOW

The fingertips of the right "5" hand are slowly drawn from fingertips to wrist of the left open hand (palm down). The slow drawing action is representative of the meaning of the word.

(264)

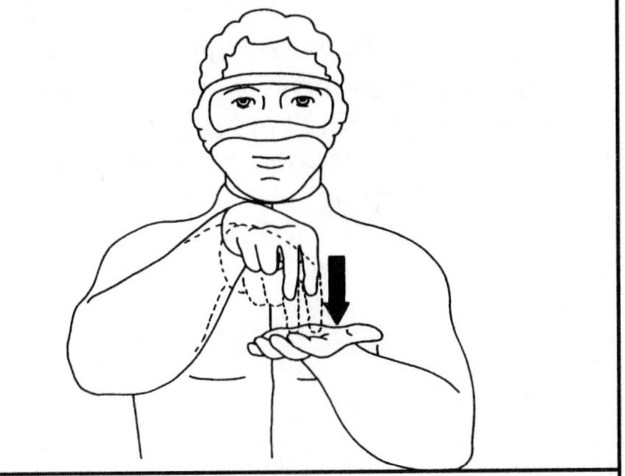

STAND

The fingertips of the right "V" hand are stood in the palm of the left open hand. The "V" fingertips represent a person's legs.

(265)

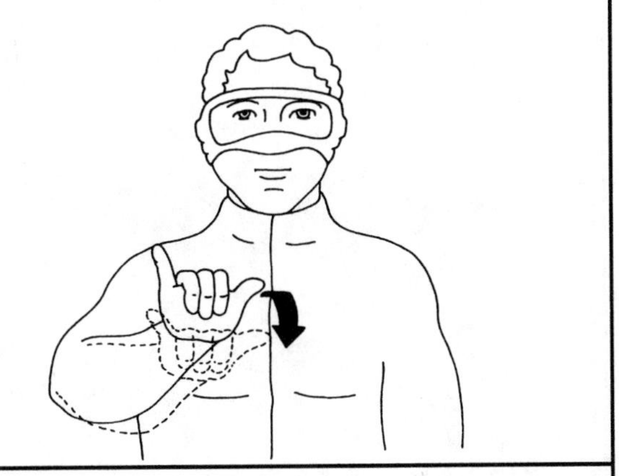

STAY

The right "Y" hand (palm down) is moved a short distance downward and outward and abruptly stopped.

(266)

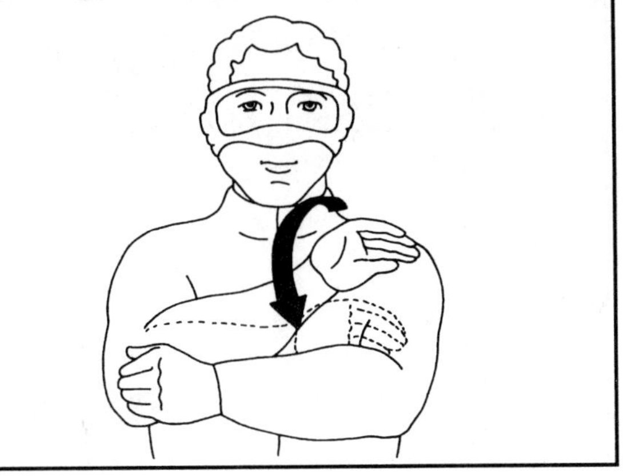

STRONG

The right open hand (palm down) is touched to the front of the left shoulder and then to the inside of the upper left forearm. This movement traces the size of the left bicep, thus indicating strength.

(267)

TALK

The index fingers (palms facing) are alternately moved back and forth from the lips/regulator indicating an exchange of conversation between two people.

(268)

TASTE

The middle finger of the "5" hand is touched lightly several times to the mouth/regulator. This movement represents a piece of food being touched to the lips for sampling.

(269)

THEY

The index finger is pointed toward the objects being described.

(270)

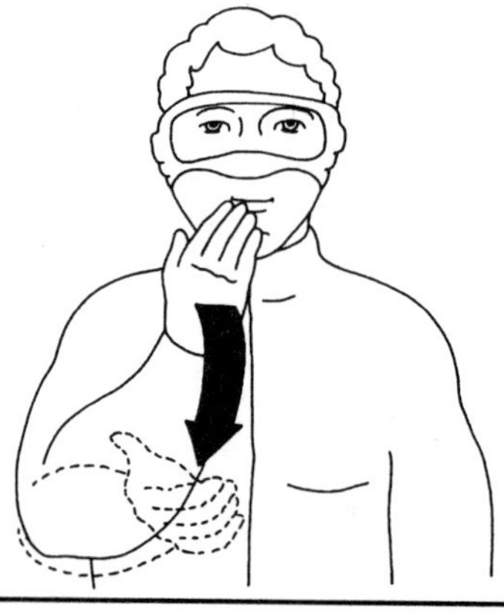

THANKS (refer to good)

This movement is basically the same movement as "good." The situation will dictate the meaning of the sign.

(271)

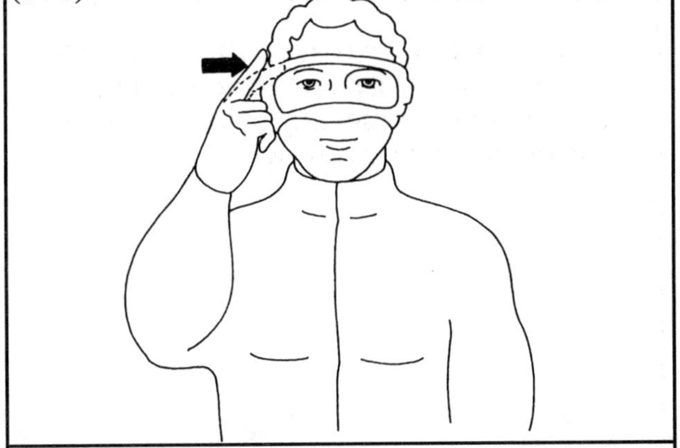

THINK

The right index finger is touched to the right forehead indicating mental action/thinking.

(272)

TOMORROW

Touch the right cheek with the right "A" hand (palm facing cheek); move the hand forward and away from the cheek with the thumb pointing forward. The action of the thumb pointing forward indicates an occurrence in the future or tomorrow.

(273)

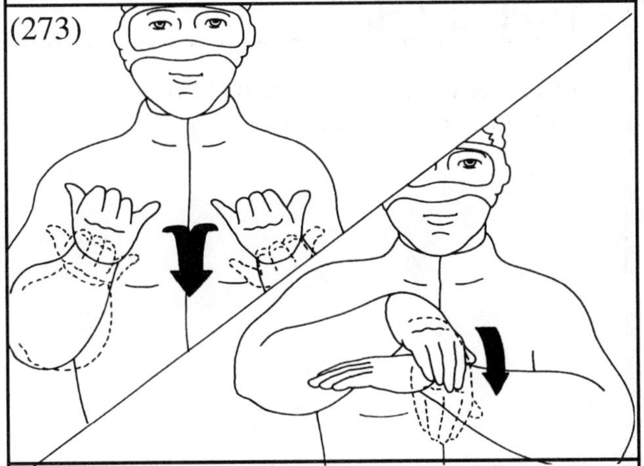

TONIGHT (dual sign)

a. To (sign now)
Both "Y" hands (palms up) in front of the diver are dropped slightly downward.
b. Night
The right open hand (palm down) is draped across the back of the left wrist (left palm down). This movement represents the sun as having gone down.

(274)

TREE

The right elbow rests on the back of the left hand as the right "5" hand (palm left) is shaken several times. This movement represents the branches of a tree being blown by the wind.

(275)

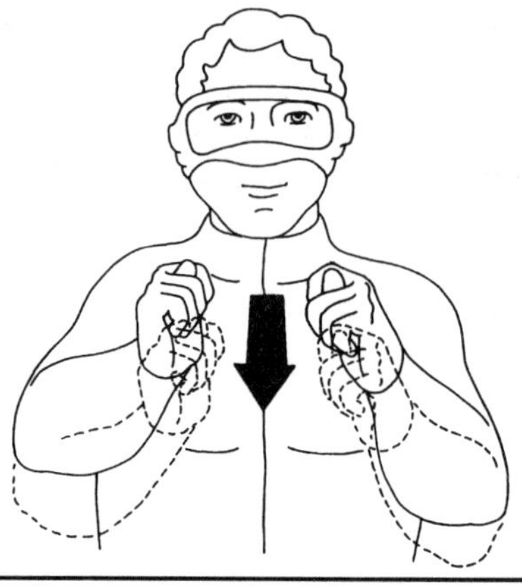

TRY / ATTEMPT

The "T" hands (palms facing) are moved simultaneously forward away from the diver. This movement represents "trying" to push a heavy object away.

(276)

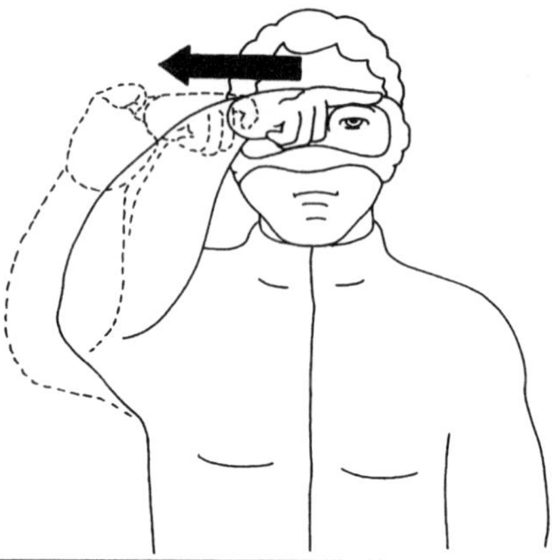

UGLY

The index finger of the right hand (palm down) is bent as it is drawn from left to right in front of the face.

(277)

UNDERSTAND

Palm facing forehead, the right index finger is flicked up. This movement indicates the "light of understanding" suddenly flashing on.

(278)

US

The right "U" hand (palm facing left) is touched to the right side of the chest and then moved in a semicircular fashion allowing the middle finger side to touch the left side of chest.

(279)

WAKE-UP / AMAZING / SURPRISE

The hands are placed on both sides of the head at eye level (index fingertips touching thumb tips; palms facing); then both index fingertips and thumbtips are snapped open. This movement represents the eyes suddenly opening.

(280)

WATER

The index fingertip of the right "W" hand is tapped on the lower lip or regulator.

(281)

WEAK

The fingertips of the right "5" hand (palm facing diver) are placed in the palm of the left open hand (palm up) and wobbled back and forth indicating "weak" knees.

(282)

WHEN

The right index finger (palm down) is circled in a clockwise manner once around the left index (pointing up) and then comes to rest on the left index fingertip.

(283)

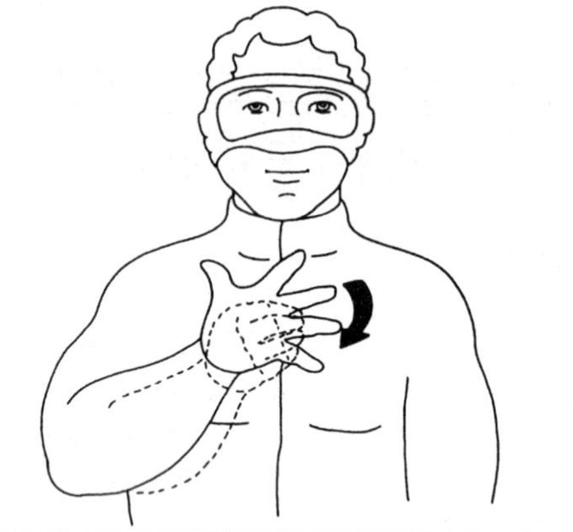

WHITE

The thumbtip and fingertips of the "5" hand are touched to the chest and then drawn forward, thumbtip and fingertips touching. This sign represents the "white" lace on a blouse or shirt.

(284)

WHY

The fingertips of the right open hand are touched to the forehead and then drawn away from and in front of the diver into a "Y" hand-shape (palm facing diver).

(285)

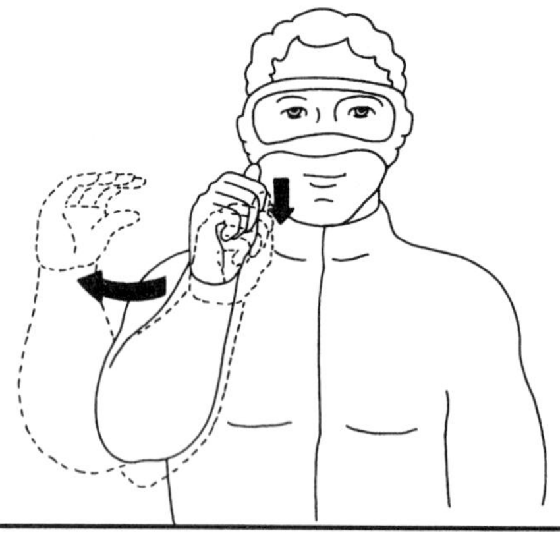

WOMAN / FEMALE

The thumbtip of the right "A" hand (palm facing left) is drawn down the right cheek. The approximate height of the female may then be indicated by placing the open hand (palm down) at a level corresponding to her height.

(286)

YOUR / YOURS

The open palm (facing out, fingers pointing up) is shoved toward the person who is being addressed.

Chapter 8

PRACTICE MAKES PERFECT

(Use of underwater communication in practical applications)

This chapter is designed in a manner that will afford the diver an opportunity to practice the signs which have been mastered in a meaningful, highly practical way. The sign translations of these English dialogues are only suggested possibilities. Each diver will of course evolve his/her own unique method of expressive underwater communication utilizing the signs presented in this text. The following practice sentences have been arranged in dialogues to facilitate continuity of thought. English sentences are presented with their "sign language" equivalents. Fingerspell the words that you do not know.

DIALOGUE 1

1. What's the name of that fish?
   (NAME + FISH + point to the fish)

2. I think it's a red snapper.
   (THINK + RS)

3. It's a beautiful fish!
   (BEAUTIFUL + FISH)

4. Yes it is; and it's good to eat.
   (YES + AND + GOOD + EAT)

DIALOGUE 2

1. What's wrong?
   (WRONG or MISTAKE and shrug shoulders)

2. Be careful, there's a moray eel.
   (CAREFUL + MORAY EEL + point to eel)

3. Where? I can't see it.
   (WHERE + I + CAN'T + SEE)

4. It's under the yellow coral.
   (UNDER + YELLOW + CORAL)

5. Does it bite?
   (BITE)?

6. It sure does and it hurts very much.
   (YES + HURT + MUCH)

DIALOGUE 3

1. Are you having fun?
   (HAVE + FUN + YOU)

2. Yes, I like it here.
   (YES + I + LIKE)

3. How deep are we now?
   (HOW + DEEP + NOW)

4. My depth gauge says 75 feet.
   (Point to depth gauge + 75)

5. Is it dangerous to go deeper?
   (DANGEROUS + GO + point to bottom + MORE)

6. No, it will be OK.
   (ALL RIGHT + GO + point to bottom)

DIALOGUE 4

1. Is it time to go up?
   (TIME + GO + point to surface)

2. No, we have five minutes left.
   (NO + HAVE + 5 + MINUTES + LEFT / REMAINING)

3. But, I feel sick.
   (I + SICK)

4. Are you dizzy?
   (DIZZY + YOU)

5. Yes, and I feel nauseated!
   (YES + DIZZY + NEED / HAVE TO + VOMIT)

6. Go slowly to the top - I will go with you.
   (SLOW + point to surface + I + GO + WITH + YOU)

DIALOGUE 5

1. My buoyancy compensator is ripped.
   (MY + BC + BROKEN)

2. What happened?
   (WHAT + HAPPENED)

3. I don't know.
   (DON'T + KNOW)

4. Will you be all right?
   (ALL RIGHT + YOU)

5. Yes, I will be careful.
   (ALL RIGHT + ME + CAREFUL + ME)

DIALOGUE 6

1. Did you see the shark?
   (SEE + SHARK + YOU)

2. No, where is he?
   (NO + WHERE)

3. I don't know; maybe he left.
   (DON'T KNOW + MAYBE + LEFT)

4. There he is.
   (point to shark)

5. What kind is it?
   (NAME + SHARK)

6. It's a hammerhead.
   (NAME + HAMMERHEAD)

DIALOGUE 7

1. What are you looking for?
   (WHAT + LOOKING + FOR)

2. Lobsters.
   (LOBSTER)

3. There are no lobsters here.
   (LOBSTER + NOT + HERE)

4. Why not?
   (WHY)

5. The water is too cold.
   (WATER + COLD)

DIALOGUE 8

1. Where's your speargun?
   (WHERE + YOUR + SPEARGUN)

2. It's broken, I left it on the boat.
   (BROKEN + LEFT + BOAT)

3. Can it be repaired?
   (CAN + REPAIR)

4. Maybe later I will try to repair it.
   (MAYBE + LATER + TRY + REPAIR)

5. I will help you.
   (I + HELP + YOU)

DIALOGUE 9

1. How many grouper did you see?
   (GROUPER + MANY + SEE + YOU)

2. I saw two small ones and one big one.
   (2 + SMALL + 1 + BIG)

3. Show me where they are.
   (SHOW + ME + WHERE)

4. They are behind the old boat.
   (BEHIND + BOAT +point to boat)

5. I'll follow you.
   (I + FOLLOW + YOU)

6. OK, let's go.
   (ALL RIGHT + COME)

DIALOGUE 10

1. Do you need help?
   (YOU + NEED + HELP)

2. I lost my find.
   (I + LOST + FIN)

3. Where is it?
   (WHERE)

4. Under the large rock.
   (UNDER + BIG + ROCK + point to rock)

5. Wait here, I'll get it.
   (YOU + WAIT + I GET)

6. OK, be careful.
   (ALL RIGHT + CAREFUL + YOU)

INDEX

A

B

C

D

T

U

V

W

Y